International Perspectives on Aging

Series Editors:
Jason L. Powell
Sheying Chen

For further volumes:
http://www.springer.com/series/8818

Donald T. Rowland

Population Aging

The Transformation of Societies

Springer

Donald T. Rowland
School of Sociology
Research School of Social Sciences
The Australian National University
Canberra, A.C.T. 0200, Australia

ISBN 978-94-007-4049-5 ISBN 978-94-007-4050-1 (eBook)
DOI 10.1007/978-94-007-4050-1
Springer Dordrecht Heidelberg New York London

Library of Congress Control Number: 2012937475

Printed on acid-free paper

Springer is part of Springer Science+Business Media (www.springer.com)

To Jennifer, Adele and Madeleine

Preface

Population aging is one the main processes transforming contemporary societies. Although it has so far proceeded slowly and inconspicuously its consequences are substantial and cumulative. Some countries already have significant experience in responding to population aging, but others do not and they face real issues in relation to demographic sustainability, fiscal sustainability and appropriate provision for the welfare of older people. Foresight and resilience-building are necessary in negotiating the transition to an older population. The Global Financial Crisis and rising government debt clearly demonstrated the repercussions of lack of foresight about emerging risks and lack of resilience in the face of adversity.

This book presents an overview of the causes, consequences and policy implications of population aging. It provides a foundation for understanding and reflecting on major demographic and social trends, together with related theoretical and policy frameworks that are important in explaining changes and designing informed responses. Drawing examples from developed and developing countries, this book presents a synthesis of recent research in the field, new analyses of trends and a discussion of the major social policy strategies. It brings together material, essential for grounding in the field of population aging, which has previously been widely dispersed. International comparisons show the prevalence of issues and the extent of differences in past and prospective experiences of population aging.

The comparisons refer especially to 38 selected countries, with populations of one million or more in 2000. They consist of 26 countries with the oldest populations projected for 2025 (20% or more aged 65 and over), eight countries with the largest aged populations in 2025 (10 million or more aged 65 and over) and a further four countries of particular interest for comparisons: Australia and New Zealand, together with the Russian Federation and Ukraine – the most populous countries in Eastern Europe. Regional comparisons refer to the United Nations' classification of regions, including its distinction between countries in 'developed' and 'developing' regions. The latter dichotomy is but a first approximation to regional variations.

Structure and Contents

To provide an overview of population aging and its implications the book is divided into three parts which examine trends and impacts of population aging at different scales. This is because population aging has wide ramifications – affecting whole societies as well as communities within them and the lives of families and individuals. There are also interactions between processes at different levels of society. For example, the future of national population aging is closely tied to birth rates, which reflect the family-building decisions of couples. Similarly, demands upon the health systems of aging societies depend substantially on individuals' abilities to preserve their own health.

- Part I (Chaps. 1–5) is concerned with developments at the societal level.
- Part II (Chaps. 6–11) focuses, in turn, on communities, families and individuals.
- Part III (Chaps. 12–17) discusses policy concepts and initiatives formulated at global and regional levels, together with prospects for national populations.

Part I: Aging and Societies

The first part of the book covers the main theories and ideas that underpin explanations and forecasts of population aging, comparisons of national trends, and changes in population health – a prime concern in aging societies. Chapter 1 (*A Silent Revolution*) introduces the emerging revolution in national age structures and the main demographic theories that have served as starting points for interpreting population aging and its consequences, namely the demographic transition, the second demographic transition and the epidemiologic transition.

This sets the scene for a comparison, in Chap. 2 (*The New Demography*) of the old demography of aging, derived from classical transition theory, and 'the new demography', based on current theories and observations. The new demography has produced a sharply contrasting set of expectations about the future of population aging in many countries, consequences of which Chap. 2 examines through models of the effects on age structures of likely rates of fertility, mortality and migration. The models illustrate possible changes in the size and structure of populations, as well as the viability of policy proposals, such as ideas about using immigration to counter aging.

Chapter 3 (*National Trends*) then presents historical background on population aging, followed by an analysis of the main patterns of age structure evolution from 1950 to 2025 in regional and national populations. This reveals the marked impact of the new demography of aging. Later sections are concerned with associated labour force trends and generational shifts affecting the ability of societies to maintain their populations, such as rises in dependency ratios and the tendency for the aged to become more numerous than children.

A major aspect of aging and societal change is the progress of the epidemiologic transition, which is concerned with trends in mortality and health. Chapter 4 (*The 'New Era' in Health*) reviews theories and ideas relevant to explaining the current stage of this transition and key features of mortality and morbidity in aging populations. This leads into Chap. 5 (*Survival and Health*) which compares societies in terms of indicators of population health, including expectations of healthy life, the duration of ill health at older ages and the causes of lost years of healthy life.

Part II: Communities, Families and Individuals

Part II examines impacts and implications of population aging at smaller scales, where the effects of changes are often most immediate and apparent, and where decisions and actions can also contribute to shaping the future of whole societies. Chapter 6 (*Community Change*) focuses on the main processes of population aging in communities – aging in place and migration. The chapter reviews explanations of these processes as well as their consequences for population composition. Amid the changes, an important question is: how can older residents thrive and contribute to their communities? Chapter 7 (*Community Participation*) addresses this, especially through an exploration of the concepts of social integration and social capital.

Family change is at the centre of the forces responsible not only for population aging, but also for shifts in the personal resources that have long been vital to the well-being and support of older people. Chapter 8 (*Family Change*) provides an overview of the importance of the family in later life, together with recent changes in the family that affect population aging and older people's support networks. The second demographic transition serves as the starting point for explaining contemporary developments. The next chapter, (Chap. 9 *Family Resources*), is concerned with life course events (e.g. marriage, divorce and widowhood) and changes in living arrangements that have a bearing upon people's resources for maintaining independence, security and social integration in later life.

At the individual level there is considerable potential for lessening some of the social and personal costs of population aging. Chapters 10 and 11 examine this with reference to two major concepts, Rowe and Kahn's 'successful aging' and Laslett's 'Third Age'. Chapter 10 (*Successful Aging*) is concerned with a research-based and policy-relevant concept that has offered a new positive direction for the study of aging at the individual level, superseding the earlier emphasis on decline and loss. The chapter explains the origins and nature of successful aging, together with its limits. Chapter 11 (*The Third Age*) focuses on the characteristics of a new stage of later life which, like successful aging, illustrates past and prospective developments in older people's experiences and the potential benefits for individuals and societies that may ensue. Chapters 10 and 11 are closely related in their subject matter.

Part III: Policies and Prospects

Part III is concerned with the outlook for countries with the oldest, or largest, aged populations and with approaches to social policies for aging societies – as formulated at international forums. Policies and prospects in Asia's developing countries receive particular attention, balancing the emphasis on developed countries in the rest of the book. Chapter 12 (*Policy Concepts*) discusses policy concepts for aging populations, such as active aging, positive aging and human capital. They are important in the academic literature and in the strategies of some national governments and international organizations. Their rise reflects the challenges inherent in the new demography of aging and a deepening understanding of the consequences of aging for individuals and societies. Although the concepts vary in meaning and currency, together they go far towards encompassing goals and focusing discussion and debate. They demonstrate recognition of the need for substantial policy agendas to address the consequences of individual and population aging.

Chapter 13 (*Policy Responses*) analyses the *Political Declaration* and the *International Plan of Action* from the Second World Assembly on Aging (Madrid 2002), which aimed to establish priorities and practical guidelines for policies in developed and developing countries. In evaluating these documents, the chapter discusses the virtual absence of policy concepts in them and the lack of attention to demographic concerns. It also examines prospects for population policies to address the latter.

The next two chapters focus on trends and policies in Asia. Much of the numerical growth of the world's aged population is occurring in developing countries, with Asia having the greatest share because of the presence of its demographic giants – China and India. Chapter 14 (*Aging in Asia*) begins with the debate about whether population aging is a significant concern for developing countries. It then reviews population trends and projections in Asia, including the emergence of the 'demographic dividend' – arising from labour force growth in some countries – and issues relating to the status of the aged and dependency levels. This establishes a setting for the discussion, in Chap. 15 (*Policy Responses in Asia*), of Asia's international policy documents on aging, especially the Macau Plan of Action (1999) and the Shanghai Implementation Strategy (2002). Because of the wide-ranging nature of these documents, the chapter argues for attention to priority areas for low income countries, especially poverty reduction, health care and support for families.

The concluding chapters review prospects for aging populations, broadly in Chap. 16 then in more detail in Chap. 17 with reference to the 38 countries specially selected for comparison. Chapter 16 (*Prospects*) focuses on overall prospects for aging societies, through a consideration of general theories of long-term change and initiatives conducive to demographic and fiscal sustainability – including the role of the Third Sector. Finally, Chap. 17 (*Risks and Resilience*) compares countries in terms of the 'risks' they face in population aging together with their potential resilience as changes occur. The comparisons provide a summary perspective, showing that societies differ considerably in terms of their likely exposure to population aging and their resources for dealing with it.

Overall, this book aims to present a broad, interdisciplinary perspective on the determinants and consequences of population aging. This is consistent with demography's approach to the study of aging. Ever since its foundation in the work of John Graunt (1620–1674) demography's core concern has been the advancement of human welfare. In pursuing this, demography has developed distinctive methods and concepts to explore the nature and implications of population trends. At the same time demographers have drawn upon and contributed to research in other disciplines, such as sociology, geography and economics, to explain processes of change and clarify consequences. Such ventures have been essential not only to theory development, but also to applications of research in planning and policy making. This book seeks to reflect these aspects of the study of population aging which, in the first instance, is a demographic phenomenon.

Acknowledgements

I wrote this book in the School of Sociology at the Australian National University. I am very grateful to the School for its support, to Madeleine Rowland for reading the manuscript, and to colleagues and students who have aided my research and teaching in demography and aging over many years.

Canberra
November 2011

Contents

Part I
Aging and Societies

Chapter 1
A Silent Revolution

... we are in the midst of a silent revolution. It is a revolution that extends well beyond demographics, with major economic, social, cultural, psychological and spiritual implications.

(United Nations Secretary-General Kofi Annan, at the launching of the United Nations International Year of Older Persons 1998)

1.1 Demographic Autumn

Social scientists have long known that world population aging would culminate in the twenty-first century, but few anticipated until relatively recently that some nations with the oldest populations could face the prospect of a 'demographic winter'. This would entail severe population decline and excessive aging, rather than a more hospitable 'demographic autumn' of population stability where the numbers of children and the elderly become nearly equal and constant. Currently the process of population aging is progressing beyond formerly perceived limits in most developed countries, with extreme developments in Central, Southern and Eastern Europe together with parts of Asia. United Nations projections for the whole of Europe envisage that, if birth rates follow current trends, its total population would decline by 70 million between 2000 and 2050, a figure greater than the present population of the United Kingdom (United Nations 2009). Although some welcome population decline as a means of restraining global warming and conserving the earth's resources, population decline heightens the effects of population aging and makes its consequences more unmanageable. Contemporary developments are undermining many national labour forces, at the same time bringing high peaks in needed expenditure on pensions, health and welfare. Occurring also is a trend towards people aged 65 years and over becoming more numerous than children and expanding their overall representation to between 25% and 35% of many national populations.

D.T. Rowland, *Population Aging: The Transformation of Societies*,
International Perspectives on Aging 3, DOI 10.1007/978-94-007-4050-1_1,

In developed countries, population aging – defined as an increase in the proportion of the population in older ages – long appeared to be a relatively benign process that would continue to develop slowly, occasion only manageable concerns, and reach its limit when about 16% of any national population was aged 65 or more. During the last quarter of the twentieth century the persistence of very low birth rates, together with ongoing improvements in survival at older ages, signaled a significant departure from past trends. Higher proportions in older ages will create a need for more thoroughgoing adaptations, for instance as pressures grow to divert more resources to older age groups. Population aging is destined to remain one of the major forces of social change in the first half of the twenty-first century and beyond. It will have consequences for young and old.

Until recently, lack of recognition of the new circumstances encouraged laissez-faire attitudes to population aging, finding reassurance in the belief that countries with the oldest populations have not been affected adversely so far. At the other extreme, awareness of the deepening of population aging has sometimes evoked alarmist claims of imminent crises for health and welfare systems. More realistically, it is clear that the challenges of population aging will not begin to peak until the 2020s. All societies will be affected, especially developed countries that already have the most rapidly aging populations, and developing countries with rapidly growing numbers of vulnerable people, young and old. In developing countries there often are meager resources for health services and social expenditure, let alone for supporting the burgeoning numbers of the elderly who have additional age-related needs. By 2050, nearly 80% of the world's older people will be living in developing countries. There will then be two 'super giant' populations of older people: China could have 331 million – almost as many as in all the developed regions combined – while India's total may be 222 million. At mid-century, India and China together could have nearly 40% of the world's aged.

There is potential to address adverse trends provided there are timely, proactive efforts to prevent them from escalating. The sudden failure of banks and other major companies in the 2008–2009 global financial and economic crisis provided a dramatic illustration of the pitfalls of ignoring risks and failing to build resilience for challenging times. Successful negotiation of population aging calls for foresight concerning prospective developments and efforts to build resilience. While population aging and growth in the numbers of older people are affecting most countries, they currently differ in the pace of change and their potential to adapt. The differences are such that regions range between being relatively favourably positioned, such as North America and Australasia, or open to adverse outcomes in the future, such as Eastern Europe and the world's developing countries.

Awareness of unfavorable prospects is a starting point for interventions. One of the chief benefits of long range population projections is that they serve, not as predictions, but as warnings of circumstances to be averted through timely action. Delays in addressing what might seem issues for future decision makers will aggravate difficulties and make them more intractable. This is because tendencies towards population decline and excessive aging steadily acquire a momentum of their own that becomes increasingly difficult to slow, let alone reverse. Negative trends become self-reinforcing if populations develop undermined age structures as successive generations beget ever lower numbers of children. The social setting for family formation is therefore crucial. Although there is no overarching theory describing and explaining population aging, there are a number of relevant theories, including transition theories, which focus on the nature of long-term changes in national populations. These help to show the progress of the “silent revolution”, elucidating changes and providing a broad setting for discussing international trends. This chapter first introduces the concept of ‘later life’ and global trends in aging as the setting for a consideration of the transition theories and their limits.

1.2 The Threshold of Later Life

Perceptions of the nature of changes and their implications depend considerably on the age-threshold adopted to describe the older population or those in ‘later life’. ‘Old age’ is a social construct with meanings that differ and change over time and from place to place. Short life expectancy, for example, is sometimes associated with an early start to old age because it is a sign of harsh living conditions, economic and social disadvantages and poor control of diseases and environmental hazards. Marked inequalities in length of life still persist within and between countries. For both sexes, life expectancy at birth ranges today from 52 years in sub-Saharan Africa, to 67 in developing countries generally and 77 in developed countries (Population Reference Bureau 2010). Longer-lived populations do not necessarily have increasingly higher perceived thresholds for the start of old age. For more than a century, 65 years remained a common pensionable age for men in industrialized countries that were experiencing the greatest improvements in life expectancy. Realization of the economic burden implied in maintaining this, however, has recently prompted a shift towards higher ages of eligibility for pensions. Extension of working life, on a full-time or part-time basis, is emerging as an economic necessity (Vaupel and Loichinger 2006).

Individual aging has biological, psychological and sociological dimensions. This results in considerable diversity in characteristics and capacities among people who are the same chronological age, which is the most comparable and only practicable general purpose measure of age. Chronological age is relevant as a partial indicator of life stage, health, labour force participation and income. It also defines generational (birth cohort) membership, which determines the historical context of individual’s lives. People born in the same year are subject to the same events and changes that can shape collective fortunes and misfortunes at each stage of life. Examples include being of military service age at the outbreak of a war, or belonging to an unusually large generation, such as a baby boom generation. During the latter’s working lives many have experienced heightened competition for employment and promotion and, as they retire, their numbers are placing increasing pressure on government financing of pensions.

Studies of population change have long employed age 65 as the start of later life. In the United States, for example, the National Research Council (2001: 30) defined 'the elderly' as persons 65 and over. Nevertheless, age 60 has been the threshold of old age in some United Nations publications, evidently in recognition of the growth in the numbers of older people in the shorter-lived populations of developing countries. Using this lower cut-off greatly augments the numbers in the older population and changes its composition. Globally, the population aged 60 and over in 2000 was 45% higher than the population 65 and over. The corresponding figure for Europe was 38% (United Nations 2009). Although there are grounds for employing the younger boundary for less developed countries, where life expectancy is lower and ill health and lifetime poverty curtail economically active life, the opposite arguments support a higher threshold for developed countries. The absence of an agreed marker for the start of later life or old age results in different findings about the course of population aging, the lower the threshold the more pronounced the apparent changes and the greater the likelihood of overstatement.

This book uses age 65 years as the start of later life because it provides comparability with many national and international studies, as well as with the demographic literature. It is not a genuine boundary of 'old age', nor is there one. Age 65 also serves as a consistent starting point for identifying stages of later life. A utilitarian approach to life stages has been the long-standing practice of distinguishing between the young-old (65–74 years), the old-old (75–84 years) and the oldest-old (85 years and over), which assumes significant differences in health statuses and needs between broad groupings, despite exceptions due to many people remaining healthy and independent throughout long lives. More satisfactory is Peter Laslett's (1989) reconceptualization of later life in contemporary developed countries as encompassing people's Third and Fourth Ages. A key distinction between them is that active living and independence characterize the Third Age, while dependency is the main characteristic of the Fourth (see Chap. 11).

Ten per cent of the population aged 65 and over represents a convenient lower limit to denote the group of countries experiencing appreciable population aging. It is a figure requiring fairly low birth rates by historical standards, that is, long-term total fertility rates of less than three children per woman. Twenty per cent is taken as the point of entry into the group of countries with the oldest populations, because 20% is close to the perceived upper limit of population aging in classical transition theory. Beyond this, 30% age 65 and over is taken as the threshold of 'hyper-aging'. It seems inevitable that figures of this order will be associated with destabilized age structures, more older people than children and adverse consequences for economies and societies. The slow unfolding of the hyper-aging scenario, however, dampens the sense of urgency and encourages procrastination.

1.3 Global and Regional Changes

For most of human history, the world's population would have had a young, pyramid-shaped age structure, with lower numbers in each older age group, which is consistent with a long run balance between high birth and death rates. At a smaller

Table 1.1 Numbers and percentages of older people in United Nations regions, 1950–2050

	Population aged 65 and over				Population aged 80 and over			
Region	1950	2000	2025	2050	1950	2000	2025	2050
Numbers (millions)								
World	131	417	832	1,487	15	70	161	395
More developed regions	64	172	266	334	8	37	71	121
Less developed regions	67	246	566	1,153	6	33	90	274
Africa	7	27	59	142	1	3	7	21
Asia	57	211	472	906	5	29	81	228
Europe	45	107	152	189	6	22	39	66
Latin America & the Caribbean	6	30	71	142	1	6	15	40
Northern America	14	40	73	98	2	10	17	36
Oceania	1	3	6	10	0	1	2	3
Percentages								
World	5.2	6.9	10.4	16.2	0.6	1.1	2.0	4.3
More developed regions	7.9	14.3	20.8	26.2	1.0	3.1	5.5	9.5
Less developed regions	3.9	5	8.4	14.6	0.4	0.7	1.3	3.5
Africa	3.3	3.3	4.2	7.1	0.3	0.4	0.5	1.1
Asia	4.1	5.7	9.9	17.3	0.4	0.8	1.7	4.4
Europe	8.2	14.8	20.8	27.4	1.1	3.0	5.3	9.6
Latin America & the Caribbean	3.5	5.8	10.6	19.5	0.4	1.1	2.2	5.5
Northern America	8.2	12.4	18.3	22.0	1.1	3.3	4.4	8.0
Oceania	7.3	9.9	14.6	18.7	1.0	2.3	3.6	6.5

Source: United Nations (2009)

scale, epidemics, famines, wars, natural disasters and migrations would have often resulted in irregular age structures. Despite this, populations with high birth and death rates have an inherent tendency to 'forget' their past and return to a young age structure. This would occur in about 60 years of unhindered natural increase. Statistics for the global population are not known accurately but United Nations estimates show the world's population with around 5% in the older ages in 1950, rising to 7% in 2000, 10% in 2025 and 16% in 2050 (Table 1.1). This major change is the net outcome of varied degrees of population aging in different regions. Figure 1.1 illustrates the prospect of considerable global aging by mid-century, as well as the inherent potential for further substantial changes later.

By 2050, the more developed regions may have 26% aged 65 and over, compared with 15% in less developed regions (Table 1.1). The latter figure reflects the likely continuation of birth rates above replacement level in many developing countries. The highest projected figure for all regions in 2050 is 31% 65 and over in Southern Europe, compared with 27% in Europe as a whole and 24% in Eastern Asia – the oldest region apart from Europe (United Nations 2009). Realization of the projection for Southern Europe could be calamitous for national economies in the region. At the same time the aged population of all regions will not only be growing larger,

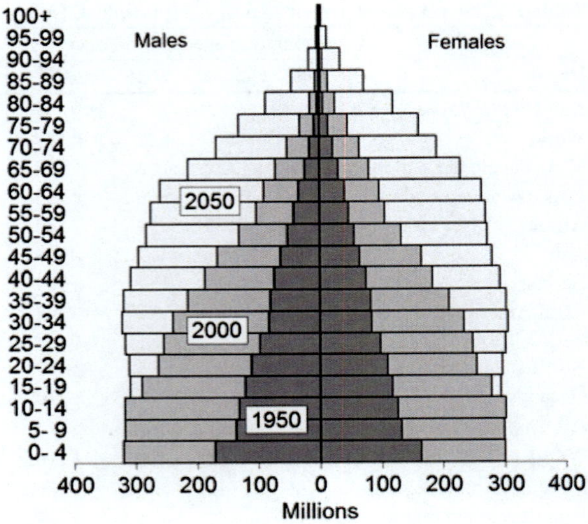

Fig. 1.1 Global age structures, 1950, 2000 and 2050 (Source: United Nations (2009), estimates and medium variant projections)

but also older. In 2050, people 80 years and over may comprise at least a third of the aged in Europe as well as in more developed regions generally, compared with about a fifth in 2000 (Table 1.1). This in itself is a considerable shift that will augment the costs of later life dependency. Even in less developed regions there will be a significant trend towards an aging of the aged.

Future numbers of the aged are better known than future percentages, because the world's aged at mid-century are already born, whereas percentages depend to a considerable extent on the future numbers of children yet to be born. In the early decades of the twenty-first century, people born in the twentieth century's population explosion, will enter the older ages, as will the developed countries' baby boom generations born in various intervals between 1946 and 1966. In the first half of this century, the numbers aged 65 and over in more developed regions is projected to nearly double, rising from 172 million to 334 million. Dwarfing even this is the almost fivefold increase in less developed regions – from 246 million in 2000 to 1.2 billion in 2050 (Table 1.1).

1.4 National Population Aging

At the national level, there is no uniform advance of countries from demographic youth to old age: the ranking of countries from oldest to youngest changes from year to year. This mainly reflects differences through time in birth rates and the size of generations reaching later life. Uncertainty surrounding future changes also leads to contrasts in the projected populations of countries from one projection series to another, especially for figures 50 years into the future. Projected population sizes are highly sensitive to assumptions about future fertility (see Bongaarts and Bulatao 1999: 516). For example, the United Nations' medium variant projection for Europe at 2050 was 691 million in the 2008 series, compared with 603 million in the 2000 series. Much of the divergence occurs after 2025 because of the cumulative effect of higher birth rates in the 2008 projections (United Nations 2001, 2009). Later projections will undoubtedly continue to modify assumptions as new developments occur.

In 1950, only nine countries – with total populations of one million or more – had at least 10% of their populations aged 65 and over and the highest figure was only 11% (France). The trend towards population aging is well illustrated, however, through the rapid changes in the number of aging populations. By the year 2000, there were 41 such countries and by 2050, they could number more than 100. This century, many smaller countries, with total populations of less than a million, will also experience substantial aging. In 1950 and 1975, all of the world's aging populations (those with 10% or more aged 65 and over) were in Europe, with the exception of the United States. By the end of the twentieth century, this was still largely true despite the many additions to the numbers of aging countries. However, the new entries did include Japan, which had emerged as having one of the world's most rapidly aging populations, together with Australia, New Zealand, Canada, and Hong Kong, now a semi-autonomous region (SAR) of China.

Beyond 2025 lie the most dramatic developments. By 2050 virtually all of the countries with the oldest populations at the start of the century are projected to have more than a quarter aged 65 and over. Eleven of these, mostly in Europe and Eastern Asia, could then be facing hyper-aging – the situation in which population aging becomes most difficult to halt or reverse. The resulting 'population implosion' is potentially as consequential for the societies concerned as the population explosion. Medium variant projections for Japan, for example, indicate a possible decline of 20% in its population in the first half of this century (United Nations 2009). Countries projected to have more than 30% of their populations in the older ages at mid-century include Italy, Greece, Portugal and Spain – in Southern Europe – together with Germany, Japan, Hong Kong and Singapore (Table 1.2). By 2050, population aging is likely to be a truly worldwide phenomenon, except in much of sub-Saharan Africa. At that time the United States, the United Kingdom, Australia and New Zealand could be among the youngest of the more developed countries if their levels of fertility and net migration remain relatively high. China, with 23% 65 and over at mid-century, would have a similar level of aging (Table 1.2). Because of the great size of China's total population, it will have a considerable impact on the extent of aging world-wide.

1.5 Aging and Demographic Transition Theory

The principal framework for studying trends in population aging has been demographic transition theory (Weeks 2002: 99–106; Casterline 2003). Its relevance, at least in its 'classical' form, as conceived by its founders and based on Europe's

Table 1.2 Percentages aged 65 years and over in the world's oldest populations, 1950–2050

	Oldest populations in 2050				Selected comparisons		
Country	2010	2025	2050	Country	2010	2025	2050
Japan	22.6	29.7	37.8	Ukraine	15.6	18.6	24.7
Italy	20.4	24.4	33.3	Sweden	18.3	21.7	24.1
Singapore	10.2	22.9	32.6	Norway	15.0	19.4	23.8
Hong Kong	12.9	22.1	32.6	Denmark	16.7	21.3	23.8
Germany	20.5	25.1	32.5	Australia	13.9	19.1	23.8
Portugal	17.9	22.4	32.1	Russian Fed.	12.9	17.7	23.4
Spain	17.2	20.4	31.8	China	8.2	13.4	23.3
Greece	18.3	22.4	31.3	New Zealand	13.0	18.1	23.2
Bulgaria	17.6	21.9	30.3	United Kingdom	16.6	19.4	22.9
Slovenia	16.4	22.4	30.2	United States	13.0	18.1	21.6
Poland	13.5	21.0	29.9	India	4.9	7.3	13.7
Austria	17.6	22.0	29.4				
Croatia	17.3	22.2	28.2				
Czech Republic	15.2	20.5	27.6				
France	17.0	22.6	26.9				
Belgium	17.4	22.2	26.6				
Hungary	16.4	20.3	26.2				
Switzerland	17.3	21.9	26.0				
Latvia	17.4	19.5	25.9				
Finland	17.2	23.9	25.9				
Netherlands	15.4	21.7	25.6				
Canada	14.1	20.5	25.5				

Source: United Nations (2009)

historical experience, has waned sharply since the 1970s. Nevertheless the theory still has broad relevance in understanding the antecedents of the present situation in developed countries, as well as current trends in developing countries. It describes and explains the shift from a pre-transition stage, characterized by young triangular age structures and high birth and death rates, to a hypothetical post-transition stage with older, more rectangular age structures and low birth and death rates. In between is the transition itself when population growth and aging occur and birth and death rates drop from high to low levels.

Generally, mortality decline begins first, initiating a period of sustained growth due to births outnumbering deaths. Population aging commences when birth rates start to fall, reducing the representation of children in the population and raising the representation of adults. Fertility decline is the main cause of population aging in the classical demographic transition. In theory, the transition ends when crude birth and death rates converge again at a low level, producing a rectangular-shaped age structure with similar numbers in all age groups below the main ages of death. The final age structure is that of a stationary population with a zero growth rate, equal numbers of births and deaths each year, and constant numbers and percentages in each age group. The demographic transition began in different parts of Europe in

Table 1.3 Characteristics of populations during and after the demographic transition

	Pre-transition	Transitional	Post-transition	Future declining[a]
Annual growth rate% (both sexes)	0	3	0	–0.3
Age structure % (both sexes)				
0–14	36.9	45.4	20.1	13.4
15–64	60.6	52.1	63.9	53.3
65+	2.5	2.5	16.0	33.3
Total	100.0	100.0	100.0	100.0
Dependency ratios (both sexes)				
Child[b]	60.9	87.3	31.5	25.3
Aged[c]	4.2	4.7	25.1	62.4
Total	65.1	92.0	56.6	87.7
Percentage surviving (females)				
To age 5	46.8	81.7	98.2	99.9
To age 65	7.8	43.3	83.1	98.2
Life expectancy (females)				
At birth	20.0	50.0	75.0	88.4
At age 5	36.6	55.9	71.4	83.4
At age 65	7.5	11.9	15.7	24.0

Sources: Hauser (1976); Coale and Demeny (1983); Coale and Guo (1990: 33); United Nations (2009)

The dependency ratios refer to the total population and differ from figures based on the female population only

[a] Whereas the figures in the other columns derive from demographic models, those in the last column are based on data for Italy from 2008 medium variant projections (United Nations 2009). The rates refer to 2045–2050, other data to 2050. The percentages surviving at ages 5 and 65, and the corresponding life expectancies were estimated from model data

[b] Child dependency ratio: 0–14/15–64 × 100

[c] Aged dependency ratio: 65+/15–64 × 100

the late eighteenth and early nineteenth centuries. Since mortality decline typically preceded fertility decline, populations would have initially become demographically younger. This was because the greatest improvements in survival occurred among infants and children through better understanding of hygiene and progress in controlling infectious diseases and famines. Statistical models of the classical transition show that through these early changes the percentage of children in a population rises appreciably while the percentage in the older ages varies little (Table 1.3).

The slowness of mortality decline dampened the pace of population rejuvenation in Europe. In contrast, rapid mortality decline occurred in many developing countries from the late 1940s onwards, such that in the space of even a single decade, some experienced the mortality reductions that had taken many decades in Europe. Their more recent and rapid pace of mortality control was founded on imported aid together with the provision of basic health services. The unexpectedly huge impact of these innovations created the population explosion, an unprecedented phase of accelerated population growth. From 1950 to 1990, annual population growth rates in the less developed regions were above 2%, a level that produces a doubling of total numbers in 35 years (United Nations 2009). The peak rate, in the late 1960s, was 2.5%. The population explosion is projected to raise the numbers of the aged in developing countries to more than half a billion by 2025 and more than a billion by 2045 (United Nations 2009).

Despite Europe's early start in the demographic transition, the aging of its population began relatively late, emerging only in the last quarter of the nineteenth century when fertility decline became prominent. Scientific recognition of population aging as a phenomenon soon followed but ongoing research interest in it did not develop until after the Second World War (Myers and Eggers 1996). An important prelude, however, was a short-lived surge of attention to population aging during the Great Depression when the United States, the United Kingdom, France, Germany and a number of other countries experienced below replacement fertility for the first time. Their birth rates soon rose again and allayed national concerns about long-term aging and depopulation, but it is notable that current issues echo those that first became prominent during the 1930s. Historical statistics for Sweden illustrate the protracted nature of population aging before the last quarter of the twentieth century. Sweden's proportion aged 65 was about 5.5% in the second half of the eighteenth century, falling slightly to 5.2% in the first half of the nineteenth century as mortality decline raised the representation of children. A figure around 5% is thought to have been typical for Western Europe in 1850 (United Nations 1973: 266–268). By 1900 the Swedish figure was 8%, rising slowly to 10% in 1950 and 18% in 2010.

Sometimes the theoretical end of the transition is taken to be when replacement fertility is reached, that is when the generation in the reproductive ages produces just sufficient children to replace themselves. Replacement takes into account children's survival to maturity: hence the higher a population's death rate, the greater the number of children needed to achieve replacement. In developed countries an average of about 2.1 children per woman is sufficient to replace the population. Population aging, however, continues far beyond the time when fertility reaches this level because larger generations, born when birth rates were higher, continue to move into older ages. The end of population aging, and the theoretical end of the transition period, occur much later, when a stationary population with a fairly rectangular age structure has emerged.

Contrary to past expectations, the anticipated post-transition period has proved illusory. Birth rates below replacement level have already plunged about half of the world's population into a new era of intensified population aging. A little-known milestone in the history of the human population was passed at the end of 2003, by which time half of the world's population was living in a country, or a sub-region of a country, with a birth rate of less than 2.1 children per woman (Wilson 2004; Wilson and Pison 2004). This included China, Brazil, Thailand and about 200 million people in regions of India. Below replacement fertility will cause many countries to confront a continuing struggle for demographic viability, rather than the demographic tranquility, or demographic autumn, envisaged for the post-transition period. Developments unforeseen in classical demographic transition theory will have the greatest importance for the future of aging societies. Because the post-transition stage has not eventuated, population aging will continue much longer than anticipated and in some countries it will far exceed previously supposed limits of change.

1.6 Further Transition Theories

By the last quarter of the twentieth century, there was widespread recognition that the classical transition was often inconsistent with continuing developments. This led to the theory of the second demographic transition which aims to describe and explain family building behaviour in contemporary Europe and, by extension, circumstances in a number of other low fertility societies (Lesthaeghe and van de Kaa 1986; van de Kaa 1987; Lesthaeghe and Surkyn 2008). The theory's main preoccupation is below replacement fertility. Hence it is relevant to explaining key features of the experience of population aging in many countries since the 1970s. Adoption of the name 'the second demographic transition' (SDT) reflects that the theory's proponents regard below replacement fertility, not as part of the first or classical demographic transition (FDT), but as a key distinguishing feature of a new set of circumstances. Whereas the classical or first demographic transition anticipated an ultimate birth rate sufficient to maintain a balance of births and deaths indefinitely in the post-transition stage, the birth rate in the majority of developed countries has tended to decline even further, creating a long-term prospect of deaths exceeding births.

In European countries, van de Kaa (1987: 8–11) suggested, there was a typical sequence of events leading to below replacement fertility namely:

- A shift from legal marriage to cohabitation: van de Kaa estimated that ultimately 40% or more of men and women in Western Europe would never be legally married.
- A shift in the focus of the family from children to the adult couple. This brings greater emphasis on childbearing as a means of enriching the lives of the parents, or on childlessness as an alternative basis for achieving personal fulfilment. It marks the end of the reign, during the first demographic transition, of the 'child-king', and the succession of the 'king-pair' (van de Kaa 1987; Ariès 1980: 649).
- A shift from preventive contraception, to contraception to permit self-fulfilling choices. Contraception thus changes from a means of preventing births that could reduce a family's well-being and standard of living, to a means of achieving greater self-fulfilment, even through having no children (van de Kaa 1987: 26).
- A shift from uniform to diversified families and households. The wider spectrum of socially sanctioned choices, together with a higher incidence of divorce, leads to a range of alternatives to the nuclear family household.

Even more important than the above changes are contrasts between the attitudes that are thought to underlie the first and second demographic transitions. During the first transition, fertility control was a response to large family size becoming a social handicap, disadvantaging parents in their goals of giving children better opportunities for education and employment. Norms and attitudes were dominated by concerns for the welfare and future prospects of offspring. At the same time, secularization reduced the influence of traditional religious teachings and made more couples willing to use methods of family planning. Within marriage, the number of children was controlled – quality replaced quantity. Thus society was child-oriented – altruism was the underlying motivation in family life (van de Kaa 1987: 5). Similarly, Ariès considered that the historical decline of the birth rate, in the first demographic transition, was unleashed by an enormous sentimental and financial investment in the child. Wise management required reducing family size so that more time and care could be devoted to each child with better results: "seeing that one's children got ahead in a climate of social mobility was the deep motivation behind birth control" (Ariès 1980: 647).

In contrast, van de Kaa described the motivation underlying the second demographic transition as individualism: norms and attitudes emphasizing the rights and self-fulfilment of individuals. Others have also emphasized the importance of individualism in explaining contemporary low birth rates in more developed countries. Couples and individuals are no longer seen as planning life in terms of the child and his or her future. The child has not disappeared from such plans, but fits into them as one of the options that make it possible for adults to achieve self-fulfilment (Ariès 1980: 650). Thus the child is no longer the essential variable in plans for the future. Whereas people planned their future in terms of familism, a family-oriented lifestyle, during the first demographic transition, in the second they plan their future in terms of any combination of familism, consumerism, careerism and other lifestyles. Replacement fertility becomes unattainable when many remain single, or married and childless, or have small families in which the total numbers of children are insufficient to counterbalance the childlessness of others.

Overall, the second demographic transition argues for a turning point in demographic history entailing a shift from altruism to a greater influence of individualism, and a shift from replacement to below replacement fertility. With its emphasis on preoccupations moving from basic to higher order needs and self actualization (see Chap. 8), it has affinities with Maslow's (1954) hierarchy of needs. Although there is no consensus that current developments are best described and explained in terms of a second demographic transition, it is a significant attempt to find order in a new and diverse situation. Because it addresses shortcomings of the first demographic transition and seeks to explain below replacement fertility it is an important approach to explaining current trends in population aging. An elaboration and critique of this theory and other theories of contemporary fertility change is presented in Chap. 8.

The last column of Table 1.3 illustrates a possible long-term outcome of the second demographic transition, using projected data for Italy in 2050. Comparing the data in this column with the adjacent column for a population in the post-transition stage of the classical transition much greater population aging is apparent: Italy's population growth rate is negative instead of zero, persons 65 and over comprise 33% of the total instead of 16% and there are 62 of them per 100 persons of working age (15–64 years) instead of 25. Table 1.3 also indicates the greater life expectancy in Italy's 2050 population – 88 years for females compared with 75 years at the end of the classical transition.

Further modifications of classical transition theory explain longer survival as well as its implications for population aging and the experience of later life. Important here as a starting point is Omran's (1971, 1981) formulation of the epidemiologic transition, which is concerned with variations in countries' experience of mortality changes through time. It expands the account, in demographic transition theory, of the course of change in the occurrence of diseases and death. Other authors' later revisions to the epidemiologic transition have focused on new developments. In its original form, the epidemiologic transition was an elaboration of the changes in death rates and disease patterns associated with the demographic transition. It identified three stages, corresponding to the three main stages of the demographic transition and differentiated principally according to rates and causes of death, namely:

- Stage 1 (pre-transition): *The age of pestilence and famine*. In this initial stage mortality is high and fluctuating, preventing sustained population growth. Up to 50% of infants and children die before their fifth birthday and life expectancy at birth varies between 20 and 40 years. The representation of the aged is low because life expectancy is short and numbers decline rapidly from age to age. No more than 3% of the population reached their 65th birthday in the pre-transition period. This stage has dominated most of human history. In the United States, for instance, its passing dates from about 1875 (Rogers and Hackenberg 1987: 234).
- Stage 2 (transition): *The age of receding pandemics*. Great changes occur in this middle stage during which mortality declines progressively as epidemics decrease in frequency and magnitude. Average life expectancy at birth rises to 55 years. As the gap between birth and death rates widens, higher rates of population growth occur. This stage began in many developed countries in the nineteenth century and in developing countries after the Second World War. It characterized the period 1875–1930 in the United States.
- Stage 3 (post-transition): *The age of degenerative and man-made diseases*. In this final stage, most deaths are due to so-called 'diseases of old age', the overall death rate converges with the birth rate, and life expectancy at birth increases to little more than 70 years. This is the stage that developed countries had reached by the 1970s: further substantial progress was not expected. Indeed, in Australia at that time, life expectancy was described as 'the most stable element of the demographic scene' (Borrie 1978: 19).

Thus the three-stage model envisaged an end-point at which there was a stabilization of causes of mortality and patterns of survival. Supporting evidence was that: (i) the biblical life span of 'three score years and ten' had scarcely been surpassed; (ii) improvements in life expectancy at birth had been due predominantly to the reduction in mortality at younger ages; (iii) there had been a shift from controllable infectious and parasitic diseases to chronic diseases that are difficult to treat. As in demographic transition theory, Omran considered that the third stage of the epidemiologic transition represented a new equilibrium where the limits to mortality decline were reached. He believed that improvements in mortality from some causes, notably cardiovascular diseases, would be offset by higher mortality from other more intractable causes, such as cancers, leading to little net gain in life expectancy (Omran 1981: 174).

Omran recognized four main patterns or models of mortality change through time, each with the above three stages. The patterns differed according to the timing, speed and extent of progress through the three stages. The two main patterns were the 'Classical or Western Model' and the 'Delayed Model'. The former denoted the experience of Western societies over the last two centuries where there occurred a gradual decline in mortality in response to social, economic and environmental improvements. In contrast the Delayed Model, characteristic of many developing countries after the Second World War, entails rapid falls in death rates accomplished through primary health care, modern medical technology and international aid. These gains, however, remain incomplete without social development and greater improvements in community-based health care.

A feature of the epidemiologic transition is the trend towards an increasing concentration of deaths in older ages. Omran's framework predates recognition of the ongoing revolution in survival which is transforming the outlook for growth in the numbers and percentages of older people in national populations. These developments have led to a reconsideration of the characteristics of the current stage of the epidemiologic transition and proposals for a new fourth stage. This contrasts with trends in fertility which have stimulated proposals for a new transition rather than just a new stage. Nevertheless, many significant changes are now encompassed within the concept of the fourth stage. Chapters 4 and 5 discuss current and prospective developments in health and survival as they relate to aging and the aged.

1.7 Conclusion

Just as current trends in population aging are transforming contemporary societies well beyond previous expectations, so too research is transforming the theoretical bases of our understanding of this phenomenon. Much uncertainty remains and it is important that identifying and explaining developments in aging societies continue to receive close attention. The first half of the twenty-first century will be the most consequential period in the history of population aging. Despite uncertainty, some broad directions and policy issues are clear. Prospects include adverse trends that are potentially modifiable through timely interventions. In many countries a continuation of present demographic trends is likely to prove economically unsustainable and socially disruptive. Action or inaction before the 2020s will be decisive in averting or compounding problems later. Aging is already one of the mainsprings of social change and it is destined to gather further impetus.

References

Annan, K. (1998). *International year of older persons*. United Nations. http://www.un.org/NewLinks/older/99/older.htm. Accessed May 2011.

Ariès, P. (1980). Two successive motivations for the declining birth rate in the West. *Population and Development Review, 6*(4), 645–650.

Bongaarts, J., & Bulatao, R. A. (1999). Completing the demographic transition. *Population and Development Review, 25*(3), 515–529.

Borrie, W.D. (Chairman). (1978). *Population and Australia: Recent demographic trends and their implications* (Supplementary report of the National Population Inquiry). Canberra: Australian Government Publishing Service.

Casterline, J. B. (2003). Demographic transition. In P. Demeny & G. McNicoll (Eds.), *Encyclopaedia of population* (pp. 210–216). New York: Macmillan Reference.

Coale, A. J., & Demeny, P. (1983). *Regional model life tables and stable populations*. New York: Academic Press.

Coale, A. J., & Guo, G. (1990). New regional model life tables at high expectation of life. *Population Index, 56*, 26–41.

Hauser, P. M. (1976). Aging and world-wide population change. In R. H. Binstock & E. Shanas (Eds.), *Handbook of aging and the social sciences* (pp. 59–86). New York: Van Nostrand Reinhold.

Laslett, P. (1989). *A fresh map of life: The emergence of the third age*. London: Weidenfeld and Nicolson.

Lesthaeghe, R., & Surkyn, J. (2008). When history moves on: The foundations and diffusion of a second demographic transition. In R. Jayakody, A. Thornton, & W. G. Axinn (Eds.), *Ideational perspectives on international family change* (pp. 81–118). New York: Lawrence Erlbaum.

Lesthaeghe, R., & van de Kaa, D. J. (1986). Twee demografische transities? In R. Lesthaeghe & D. J. van de Kaa (Eds.), *Bevolking: groei en krimp, mens en maatschappij book supplement* (pp. 9–24). Deventer: Van Loghum-Slaterus.

Maslow, A. (1954). *Motivation and personality*. New York: Harper and Row.

Myers, G. C., & Eggers, M. (1996). Demography. In J. E. Birren (Ed.), *Encyclopedia of gerontology: Age, aging and the aged*. San Diego: Academic Press.

National Research Council. (2001). *Preparing for an aging world: The case for cross-national research*. Washington, DC: National Academy Press.

Omran, A. R. (1971). The epidemiologic transition: A theory of the epidemiology of population change. *The Milbank Memorial Fund Quarterly, 49*, 509–538.

Omran, A. R. (1981). The epidemiologic transition. In J. A. Ross (Ed.), *International encyclopaedia of population* (pp. 172–175). New York: Free Press.

Population Reference Bureau. (2010). *2010 World Population Data Sheet*. Washington, DC: Population Reference Bureau.

Rogers, R. G., & Hackenberg, R. (1987). Extending epidemiologic transition theory: A new stage. *Social Biology, 34*, 234–243.

United Nations. (1973). *The determinants and consequences of population trends* (Vol. 1). New York: United Nations Department of Economic and Social Affairs, Population Studies No. 50, ST/SOA/SER. A/50.

United Nations. (2001). *World population prospects: The 2000 revision (CD, disk 1: standard set)*. New York: United Nations, Department of Economic and Social Affairs, Population Division.

United Nations. (2009). *World population prospects: The 2008 revision population database*. New York: United Nations, Department of Economic and Social Affairs, Population Division. http://esa.un.org/UNPP/index.asp?panel=2. Accessed April 2010.

van de Kaa, D. J. (1987). Europe's second demographic transition. *Population Bulletin, 42*(1), 1–59.

Vaupel, J. W., & Loichinger, E. (2006). Redistributing work in ageing Europe. *Science, 312*(5782), 1911–1913.

Weeks, J. R. (2002). *Population: An introduction to concepts and issues*. Belmont: Wadsworth/ Thomson Learning.

Wilson, C. (2004). Fertility below replacement level. *Science, 304*(5668), 207–208.

Wilson, C., & Pison, G. (2004). More than half of the global population lives where fertility is below replacement level. *Population and Societies, 405*(October), 1–4.

Chapter 2
The New Demography

Those with a concern for an environmentally sustainable future may see a gradual decline in population as a blessing rather than a threat, but the prospect of population decline implied by current fertility rates is anything but gradual, threatening to halve or even quarter the populations of some Western nations over the course of the next century.

(Castles 2004: 141)

2.1 Continuing Changes

Falling birth rates are the main reason why populations grow older because they reduce the proportion of children in a population and thereby raise the proportion of older people. Amid falling birth rates, it was often assumed that societies retained an inherent tendency to at least perpetuate themselves. However, birth rates in many countries have not stabilized at a level sufficient to maintain population numbers. Completely unforeseen was the emergence since the 1970s of below replacement fertility sustained for decades, rather than being merely a transitory phenomenon. At the same time, the formerly assumed limit of life expectancy – at a peak of around 75 years for females and 71 for males – has continued to be exceeded by widening margins in Europe, North America, Australasia and most of East Asia. Passing the supposed limit of life expectancy has also overturned the notion, from demographic transition theory, that mortality decline has only a minor role in raising the percentage of older people. In the developed countries' past, improvements in survival occurred mostly among the young. This led to overall population growth and, decades later, to greater numbers reaching older ages. The effect on the percentage aged 65 and over was minor, however. Present circumstances contrast sharply. Currently, progress in mortality control has reached a stage in developed countries where death is uncommon at young ages and many of the gains are occurring in middle and later life. This new phase in the decline of the death rate has the immediate effect of

D.T. Rowland, *Population Aging: The Transformation of Societies*, International Perspectives on Aging 3, DOI 10.1007/978-94-007-4050-1_2,

Table 2.1 The old and the new demography of aging

Old demography	New demography
% 65 years and over peaks at less than 20	% 65 years and over rises to 25–35 or more
Fertility lowest at replacement	Below replacement fertility widespread
Life expectancy peaks around 75 years	Life expectancy rises to 85 years and beyond
Mortality decline: minor role in population aging	Mortality decline: major role in population aging
Slow tempo of population aging	Tempo of population aging increasing
Aging unimportant in developing countries	Numbers of the aged a major concern in developing countries

augmenting not only the numbers surviving at older ages but also the percentages. In conjunction with the extended decline of the birth rate, the extended decline of the death rate is leading to unforeseen levels of population aging.

Linked with now disproven ideas about limits of fertility and mortality rates was the expectation, discussed in Chap. 1, that the proportion in the older ages would peak as low as 16% at the end of the demographic transition. This contributed to the view that population aging denotes a victory for goal setting and policy making. Authors have also hailed it as "a great triumph of civilization" (Notestein 1954), "one of the truest measures of progress" (Cowgill 1970: 35), and "the greatest triumph that our species has achieved" (Kirkwood 2001). The "triumph" arose from humankind's ability to control disease and unsustainable high birth rates. Population aging was a sign of these achievements because it cannot occur in societies with the negative characteristics of rapid population growth or high death rates at young ages. Thus population aging was a positive development, not only because it heralded the end of the transition from high to low birth and death rates, but also because it appeared that aging would remain at a manageable level. Even as late as 1989, it was thought that "national populations in post-industrial societies … are extremely unlikely to exhibit elderly proportions much more than … 18–22% for those aged 65 and over" (Rogers 1989: 17).

Interpreting population aging as a measure of progress further assumes that the transition ceases at a theoretical end point where age structures are rectangular and population numbers are nearly stationary. Yet birth and death rates are not so closely connected that their decline terminates at equilibrium (Barclay 1966: 133). Furthermore, envisaging that population aging and progress go together presupposes that the social and economic changes responsible for falling birth and death rates permit the accumulation of sufficient resources to provide for larger numbers of older people. Instead, the provision of adequate support for the aged is an ongoing concern in all countries whatever their birth rates and income levels: generous public provision for the aged is economically unsustainable everywhere.

Replacing the former assured and even optimistic perspective on population aging, there is now mounting awareness of the extent of the continuing revolution in demographic processes. Compared with the old demography, based on the classical demographic transition, the new demography, founded on contemporary theories and observations, poses greater challenges because it entails more extreme developments to which successful adaptations are all the more important. Table 2.1

summarizes the transformation in the outlook for aging societies due to the waning relevance of the old demography of aging and the rise of the new. This chapter focuses on the implications of major characteristics of the new demography of aging, apart from its relevance to developing countries, which is discussed in Chap. 14. Ensuing sections illustrate: (i) the shift from the old to the new demography, (ii) the implications for aging of various levels of fertility, mortality and migration, and (iii) the trend towards an acceleration of demographic aging.

2.2 Overview of Trends

Figure 2.1 provides an overview of trends in aging, including the cross-over from the old demography to the new. In the diagram, the trend in population momentum through time depicts the course of population aging during and after the demographic transition. Population aging is a consequence of factors that cause changes in momentum (Myers 1990; Rowland 1996; Kim and Schoen 1997). Population momentum measures the extent to which past trends in fertility, mortality and migration have created potential for expansion or contraction to occur in the size of age groups as a result of cohort flow – the progression of birth cohorts from one age group to the next. Population momentum expands middle and older age groups in particular and aging is a direct consequence of momentum. Thus momentum indicates the potential for growth and aging inherent in a population's age structure.

Positive momentum, or 'growth potential', is prospective growth due to larger birth cohorts, or generations, growing older (Keyfitz 1971). Conversely, negative momentum is prospective decline due to smaller cohorts growing older. Negative momentum in an age structure is usually due to below replacement fertility. Positive momentum is depicted on the left of Fig. 2.1, negative momentum on the right. If the birth rate

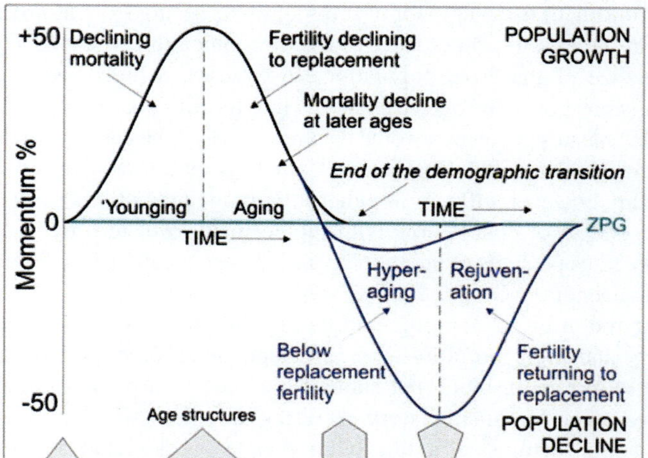

Fig. 2.1 Trends in population aging

in a growing population suddenly fell to replacement level, and the current life expectancy was maintained, the population would continue to grow for 70 years or more before levelling off (Potter et al. 1977: 555). Positive population momentum is this tendency for continued increase as younger cohorts, longer-lived than those born decades earlier, move up the age pyramid. Although no country can achieve replacement fertility instantaneously, the assumption of an immediate shift to replacement, in conjunction with the assumption of constant mortality, is useful because it permits the measurement of the current potential for change from cohort flow alone. In doing so, momentum provides an indication of how near a population is to zero population growth (ZPG) and the end of aging.

For example, in 1970, population momentum in the United States was 32%. In other words, the population had an inbuilt potential to grow by 32% in future decades through the movement of larger cohorts into older ages. This high figure reflected the considerable future growth that would arise from the progression of the country's large baby boom generation into middle and later ages (Rowland 2003: 327). Calculation of the figure of 32% assumed replacement fertility from 1970 onwards, as well as constant mortality at the 1970 level and zero migration. These assumptions eliminated sources of growth other than that built into the age structure. By 2010, the country's momentum had fallen to 7%, because the baby boom generation was 40 years older and there were fewer age groups yet to experience its impact. The fall in momentum showed that the United States was closer to ZPG. Nevertheless, a 7% increase in the size of the population, concentrated in the older ages, still constituted a substantial future change which would make zero momentum and zero growth unlikely for several decades.

Developing countries have the highest levels of momentum. For example Bangladesh, which in 2025 will have one of the world's largest aged populations, had population momentum of 40% in 2010 (Fig. 2.2). The shaded area in Fig. 2.2 denotes the size that Bangladesh's population would reach with constant replacement level fertility. The difference between the size of this population and the 2010 population (outlined) is the amount of future growth arising from momentum alone. Expansion in the middle and older ages due to momentum would be equivalent to 40% of the size of the 2010 population. In contrast, a high degree of negative momentum is present in the age structure of low fertility societies, as illustrated by Japan in 2010. Japan's population would decline by 21% if the country had replacement fertility from 2010 onwards. Heavy losses would occur at most ages over 30. Data on momentum, however, differ from other long-range projections in that they focus only on the effects of cohort flow with replacement-level fertility. Japan's actual demographic rates place its population on a trajectory to decline by 50% or more by the end of the century (Kaneko 2008).

As illustrated in Fig. 2.1, momentum is zero in the pre-transition stage because high fertility and high mortality tend to balance each other and the numbers in every age group remain much the same. This state of affairs is thought to have characterised much of human history, resulting in regions with stationary, triangular-shaped age structures with few older people, as depicted at the bottom of Fig. 2.1. The supposed equilibrium in the pre-transition stage, however, disregards

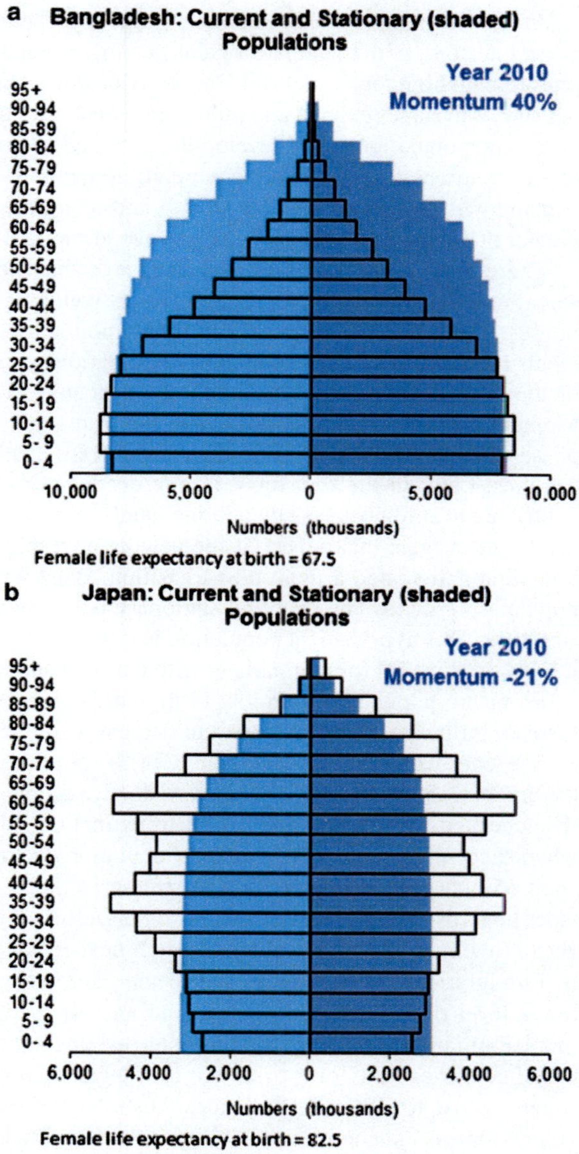

Fig. 2.2 Population momentum in Bangladesh and Japan, 2010. (**a**) **Positive momentum**: Stationary population larger than the current population. (**b**) **Negative momentum**: current population larger than the stationary population (Source: Calculated from data in United Nations 2009b)

episodes of population growth due to technological innovations and migration, as well as population decline due to the ravages of wars, natural disasters and other calamities. These would have caused multitudes of demographic upheavals and crises through time and from place to place, with a net outcome in the long run of slow population growth at best.

At the beginning of the transition stage, momentum rises above zero because death rates fall, with the greatest gains occurring among the young. Typically, birth rates remain high for a time while mortality declines. As a result, the numbers reaching successive ages expand and momentum continues to increase. During the transition stage populations may develop the potential to grow by 50% or more because of larger numbers moving up the population pyramid. The existence of momentum explains why world population growth and aging are currently unstoppable: ever greater numbers will advance from one age group to the next (see Fig. 1.1).

There is no word in the English language that means 'to make even younger something that is already young', although 'younging' and 'juvenation' are possibilities. 'Younging' occurs early in the demographic transition because improvements in the survival of children make populations younger. This broadens the base of the triangular age structure as the percentage of children increases (Fig. 2.1). Momentum keeps rising until, later in the transition, fertility starts to fall. In this phase the size of younger and older cohorts gradually evens up and momentum declines. Later in the transition, falling death rates at middle and later ages also contribute to aging because they immediately augment the number of adults surviving to older ages: this is part of the new demography. The classical demographic transition anticipated a final, post-transition, stage with zero momentum and zero population growth. The resulting stationary population has a more rectangular age structure. This hypothetical population is referred to in some later chapters because it is a useful model for comparison with national and regional age structures.

So far no population with low birth and death rates has achieved zero growth. Instead, birth rates have continued to decline below replacement level, resulting in a cross-over from the old demography of aging to the new. The right-hand side of the diagram shows the development of negative momentum and hyper-aging which, if unchecked, would ultimately lead to extinction. Momentum becomes negative when successive cohorts are smaller. Population decline ensues and the percentage aged 65 and over passes 30%. Age structures therefore become undermined and taper towards the base. Nevertheless, some Western populations could approximate zero growth around mid-century, through near-replacement fertility supplemented by immigration. This is represented by the shallower curve on the right depicting a lower level of negative momentum and an early return to ZPG. Current data on momentum in national age structures suggest this is a possibility (Fig. 2.3). A number of countries have low positive or low negative momentum – and hence are reasonably close to ZPG. This includes Australia and New Zealand, Canada and the United States, and countries in Northern and Western Europe. Developing countries mostly have high momentum, while Eastern and Southern European countries have pronounced negative momentum. Japan and Bulgaria have the lowest momentum in the data set. The deeper curve in Fig. 2.1 denotes the development of greater negative momentum. Any future recovery from this through replacement fertility would result in a much diminished population size. This is because of the severe depopulating effects of a prolonged period of negative momentum. If fertility remains well below replacement, population decline becomes self-reinforcing.

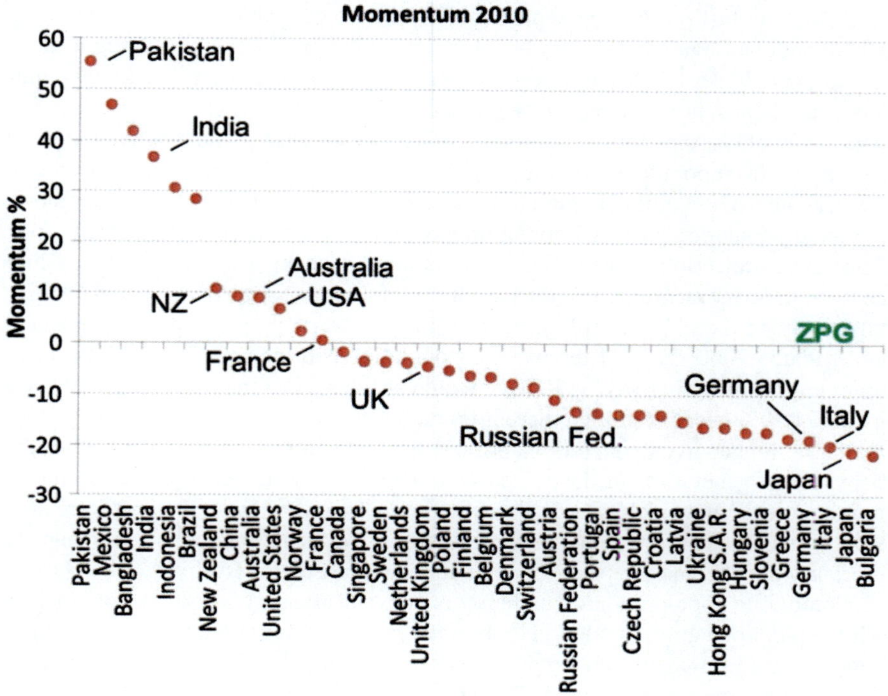

Fig. 2.3 Momentum in national populations, 2010 (Note: The statistics refer to females only) (Source: Author's calculations from population estimates (United Nations 2009b) and West model life tables (Coale and Demeny 1983; Rowland 2003))

2.3 Growth and Aging

Such an eventuality was once inconceivable, so much were concerns focused on growth and its containment. One of the achievements of twentieth century policy making, in conjunction with social and economic modernization, was progress towards reducing world population growth through enabling more women to achieve their desired number of children, rather than having little control over their own fertility. Recent examples of national fertility decline are India, Indonesia and Iran where their respective total fertility rates in 2010 were 2.6, 2.4 and 1.8 children per woman (Population Reference Bureau 2010). Unfortunately, the emergence of lower birth rates in many developing countries has helped to foster the view that further progress will follow as a matter of course. This assumes that past successes in reducing high birth rates will continue, but much will depend on the maintenance and extension of international support for reproductive health programs and social development, particularly in countries where there has been little such progress so far (Caldwell 2003: 81–82).

Achieving birth rates near replacement level are desirable goals for stabilizing global population numbers by mid-century and containing population aging. Delays in achieving lower fertility will result in much higher peaks in some national population totals and in the numbers of the aged. Nevertheless, a great increase in the population of less developed countries is inevitable because of the inherent potential for growth from population momentum as larger generations grow older. Although the persistence of medium to high birth rates in some countries is a significant factor in global population growth, population momentum is now a more important force. Illustrative projections of world population increase for the whole of the twenty-first century show momentum to be the predominant source of growth, followed by mortality decline then above replacement fertility in developing countries (Bongaarts and Bulatao 2000: 522). If the world's population had instantly achieved replacement level fertility in the year 2000, it would still have had the potential to increase by about 30% because of the inherent momentum in its age structure. The aging of the large, longer-lived generations born in the 'population explosion' together the baby boom generation in developed countries, is making huge global increases imminent and unavoidable. Global population increases of more than 50% are possible by 2050, with even greater increases in developing countries due to further mortality decline and the persistence of above replacement fertility.

Continuing population growth therefore coexists as a global issue in conjunction with population aging. In relation to the aged population, growth in their numbers is generally the immediate concern in both developed and developing countries. In the former there are expectations that policies and programs will respond quickly to changing numbers nationally and locally, and only in the longer term to structural changes due to shifts in the overall representation of older age groups. Changes in the number of people moving through the age structure also drive the nature of the market for goods and services and the demand for housing and community infrastructure. While great increases in total populations and the numbers of the aged lie ahead for many countries, overall population decline, together with labour force decline, will be difficult to avoid where fertility is very low. The mechanisms underlying the shift from high to low fertility are now reasonably well understood but still uncertain are the means of halting or reversing it when it goes too far, as is likely for example in Japan and parts of Europe. Moreover, delays in taking policy initiatives to prevent excessive population decline can postpone the time when significant benefits are possible. This is because there is a lead time in developed countries of around 30 years before most people in new generations are established in productive employment and family formation.

2.4 Contemporary Developments

While the future is necessarily speculative, many consider that below replacement fertility will persist in societies that have already reached this situation. For example, in its 2008 revision of *World Population Prospects*, the medium fertility assumption

Table 2.2 Total fertility rates in United Nations regions, 1950–2050

Region	1950–1955	1975–1980	2000–2005	2025–2030	2045–2050
World	4.92	3.83	2.67	2.21	2.02
More developed regions	2.82	1.94	1.58	1.70	1.80
Less developed regions	6.00	4.53	2.89	2.28	2.05
Africa	6.63	6.61	4.91	3.23	2.40
Asia	5.73	4.03	2.50	2.01	1.90
Eastern Asia	5.42	2.68	1.71	1.81	1.83
Europe	2.65	2.00	1.43	1.65	1.80
Eastern Europe	2.82	2.08	1.26	1.59	1.79
Northern Europe	2.38	1.85	1.69	1.84	1.85
Southern Europe	2.62	2.26	1.35	1.62	1.80
Western Europe	2.41	1.67	1.58	1.68	1.79
Latin America & the Caribbean	5.85	4.46	2.50	1.85	1.80
Northern America	3.33	1.82	1.99	1.84	1.85
Oceania	3.83	2.74	2.42	2.18	1.93

Source: United Nations (2009b), medium variant projections for 2025–2050

of the United Nations Population Division was that total fertility in all countries would eventually converge at 1.85 children per woman, although not all would reach this level before 2050 (United Nations 2009a). It is also widely expected that societies with long life expectancy will continue to experience greater longevity in coming decades. Again, United Nations projections for national populations and global regions are consistent with this. Uncertainty about population prospects increases the further into the future that projections extend. Nevertheless, there is consistency between the United Nation's recent revisions of its projections to 2050. Although the United Nations expressed the assumptions for its 2000 and 2008 medium projections somewhat differently, common to both was the expectation that below replacement fertility will persist in developed countries generally, but their total fertility in 2045–2050 will be closer to replacement than at the start of this century. After 2050, however, the 2000 series of United Nations projections assumed an eventual convergence to replacement in all regions, whereas the 2008 series assumed convergence to a figure just below replacement.

During the second half of the twentieth century, the world's estimated total fertility rate (TFR) fell by more than 40% – from 4.9 births per woman to 2.7 births (Table 2.2). This reflected the progress of the demographic transition in less developed regions together with the emergence of below replacement fertility in more developed regions. The world's oldest populations all had TFRs above replacement level in 1950–1955, including Italy (2.4) and Japan (3.0), but by the second half of the 1970s most already had below replacement fertility. Thereafter, further declines were widespread, even to TFRs of between 1.0 and 1.3. Some upturns have ensued from the lowest points. Variations between women in the timing of their first birth, the spacing of subsequent births and the total number of births create much potential for fluctuations in annual birth rates, as do external influences such as economic booms and recessions and the introduction or withdrawal of financial support for parents. Variations in birth rates occur when many women in a large cohort delay childbearing in one interval and ‘catch up’ in another.

For these reasons TFRs from annual data are only partial indicators of long-run average completed family size. Among the oldest populations, the United Kingdom, France and the Scandinavian countries have proved exceptions to birth rate decline to very low levels. Their TFRs were between 1.8 and 2.0 in 2010 (Population Reference Bureau 2010). These exceptions contribute to heterogeneity in the experience of low fertility societies over recent decades and render difficult the tasks of forecasting trends. There are also many causes of national fertility differences since national rates are an amalgam of variations in fertility between regions and social groups (Bongaarts and Bulatao 2000: 87). Regions of Africa and parts of Asia and Latin America, which have continuing high fertility, will experience rapid growth in their numbers of older people in conjunction with major increases in population size overall.

Whereas discussions of future population trends commonly refer to United Nations’ medium variant projections, low and high variant projections help to define the range of uncertainty within which birth rates may fall. For the ‘more developed regions’ the 2008 low variant projection envisages a TFR of 1.17 in the second half of the 2020s, compared with the high variant’s figure of 2.17. Although the range is only one child per woman, variations within this range are critical in determining whether fertility is above, below, or far below replacement – where its depopulating effects become extreme. The medium variant projections in Table 2.2 for 2025 and 2050 mostly envisage higher fertility than was observed in 2000, but achievement of this will be particularly difficult in countries in Eastern Asia, Eastern Europe and Southern Europe.

The effects of low fertility would be more manageable if they were not accompanied by rising life expectancies. Major gains in life expectancy occurred in conjunction with fertility decline in the second half of the twentieth century as many developing countries experienced accelerated progress through the demographic transition and most developed countries exceeded its expected limits. By the early twenty-first century developing countries had added 22 and 24 years to their 1950 male and female life expectancies, while the already long-lived populations of developed countries had added 9 years for males and 11 years for females (United Nations 2009b). The trend in developing countries gave rise to the global population explosion, and now underlies the growth momentum in many national age structures.

In contrast to the United Nations fertility projections, those for life expectancy, in both 2000 and 2008, consist of a single series and do not attempt to define a range within which future changes could occur. This is a disadvantage in view of the impacts on population aging that future changes in survival will have. Projections from national statistical agencies and demographic research organizations have been giving increasing attention to these potential variations and the next section illustrates their long-term consequences. The United Nations projections, nonetheless, provide an initial view of future prospects. By 2010, most of the world’s oldest populations had life expectancies of 76–80 for males and 82–84 for females (Population Reference Bureau 2010). For the first half of the twenty-first century, the UN's projected average gain in life expectancy at birth for the 15 oldest populations is about 5 years for both males and females. Japan, however, stands out as attaining exceptionally high female life expectancy by 2050 – 91 years, compared with 84 for Japanese males (United Nations 2009b).

2.5 Implications for Aging and Population Size

2.5.1 Effects of Fertility and Mortality

Preceding sections have shown that the extended declines in fertility and mortality have modified the conventional interpretation of the roles of fertility and mortality in population aging. The contemporary range in low fertility rates is leading to various levels of aging, all surpassing previous expectations. Extension of life at middle and older ages is reinforcing this. Demographic models reveal the full potential aging effect of long-run trends in fertility and mortality (Fig. 2.4). The models use life expectancy at birth as the measure of mortality and total fertility rates as the measure of fertility; they refer to total populations with female life expectancies between 70 and 90 years and corresponding lower male life expectancies. The lower a population's life expectancy, the higher the level of fertility needed for replacement because more die young. The total fertility rates are set at 3.0, 2.1 and 1.0, roughly corresponding to replacement levels of 150%, 100% and 50% respectively. In 2010, eight countries in Europe had total fertility rates of 1.2 or 1.3, which is little more than half the replacement level. At that time, Hong Kong, Macao and Taiwan had the world's lowest TFRs of 1.0 (Population Reference Bureau 2010).

The models show that, with replacement-level fertility, the percentage of the total population aged 65 and over reaches 18% when female life expectancy is 80 and 25% when it is 90 (Fig. 2.4). The rise, due solely to higher life expectancy, is equivalent to the effect of a substantial fall in fertility. When high life expectancy is combined with below replacement fertility, the outcome is very high levels of aging. Thus, with an average of one child per woman, the representation of persons 65 years and over reaches 38% when female life expectancy is 80, and 50% when it is 90. Clearly, extended longevity is a major force in population aging. Variously termed 'excessive aging', 'super aging', or 'hyper-aging', very high proportions in older ages represent a potential future for a number of long-lived low fertility societies, unless there are effective policy initiatives or as yet unanticipated social changes. The more sustainable scenarios in Fig. 2.4 are those where there is replacement level fertility, although even replacement fertility, in conjunction with a life expectancy of 90, results in 25% aged 65 and over.

Another concern is that TFRs well below replacement have powerful depopulating effects. This is illustrated in Fig. 2.5 where, for various low fertility rates, are

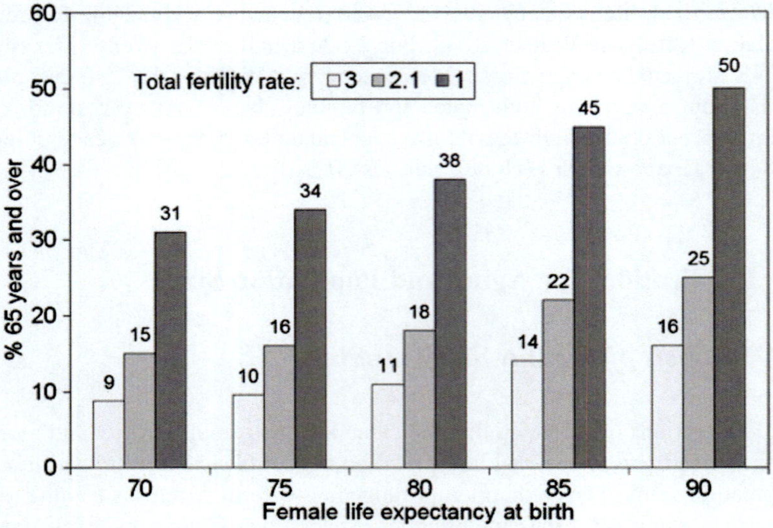

Fig. 2.4 Effects of low fertility and mortality on population aging (Sources: West model life tables, a relational model life table for a female life expectancy of 90 and stable population models: Coale and Demeny 1983; Rowland 2003: 316–321)

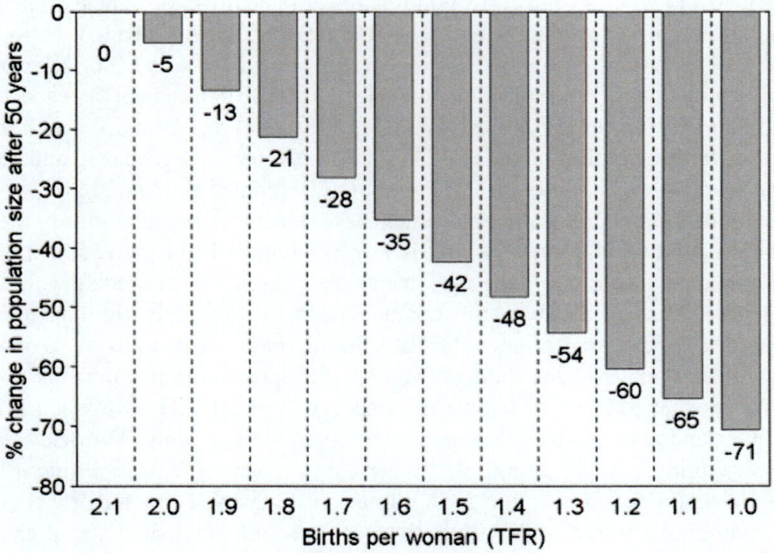

Fig. 2.5 Effects of low fertility and mortality on population size after 50 years (Note: The initial population for each projection is a stable population with the total fertility rate (*TFR*) indicated for each column. The models assume life expectancies at birth of 80 for females and 77 for males) (Sources: West model life tables: Coale and Demeny 1983; Rowland 2003: 316–319)

shown the resulting falls in population size after 50 years. The calculations for each column of the graph assume that the population is stable, that is, it has constant fertility and mortality rates through time as well as a constant age structure. The measure of fertility is again the total fertility rate, as plotted on the horizontal axis. The death rates for all the columns are based on life expectancies at birth of 80 for females and 77 for males – which approximated the average figures for Western Europe in 2009. The graph indicates that when the TFR is 2.1, no change occurs in population size though time. In contrast, TFRs of 1.6 or less bring rapid changes in population size, with numbers falling in 50 years by more than a third. The consequences for population size of even lower fertility are extreme. For example, maintenance of TFRs of 1.3, which occurred in parts of Eastern and Southern Europe in 2010, would cause a 54% fall in population size within 50 years.

Although the model populations in Fig. 2.5 are artificial constructs, they are relevant to national situations not only to forewarn of unfavourable circumstances, but also as indicators of trajectories that some countries are already embarked upon. An ameliorating factor for contemporary populations is that their current age structures are younger than those of most of the model populations. This will restrain population decline for some time, although substantial changes remain possible. For example, if Italy and Japan maintained their low mortality rates and TFRs of 1.3 their total populations would fall by 30% between 2000 and 2050. Bongaarts' and Bulatao's (2000: 6) projection for Italy also envisaged a 30% decline. Taking a TFR of 1.6 as "the benchmark for potential crisis", Castles (2004: 165) concluded that "roughly half the countries of the OECD are in deep trouble". He added that: "The evidence is not yet available to demonstrate the existence of a general crisis; only that some countries face huge problems if things stay the way they are." (ibid: 166).

2.5.2 *Effects of Migration*

Low birth rates have inevitably stimulated thinking about the ability of immigration to counter demographic decline and population aging, especially since the United Nations Population Division published its report on *Replacement Migration* (United Nations 2000). Although immigration is now recognized as no panacea for huge emerging demographic deficits, it remains one of a number of strategies with potential benefits for aging populations (Castles and Miller 2003: 82). Countries, such as the United States and Australia, which have long histories of utilizing immigration to boost labour force growth and address deficits in the young working ages, are favourably placed to do so in the future for several reasons. First, they have mostly encouraged permanent settlement, which heightens interest in these countries as destinations for family-oriented migrants. Permanent settlement also fosters the building of new generations, which supplements the numerical gains ensuing from migration. Second, they have become accustomed to integrating immigrants, granting citizenship and supporting, or at least accommodating, cultural diversity.

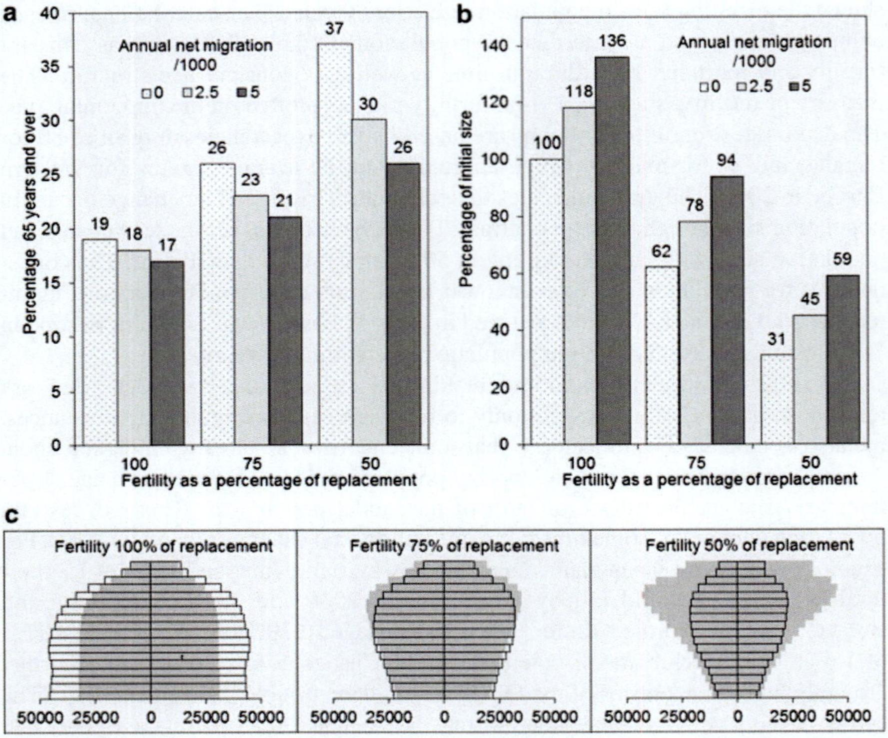

Fig. 2.6 Effects of migration on aging and population size. (**a**) Percentages aged 65 and over after 50 years of net migration. (**b**) Population size after 50 years of net migration. (**c**) Population before (shaded) and after (outlined) 50 years of net migration at 5/1,000 (Sources: Projections based on West model life tables and stable populations, see Fig. 2.5)

In contrast, temporary 'guest workers' are more likely to experience marginalization and lack of acceptance in the host society. Third, they have had long experience in recruiting immigrants in a world where the pool of tertiary-educated and occupationally-skilled potential immigrants is limited. Varying levels of interest in using migration as a policy instrument, together with competition for skilled migrants, will continue to result in an uneven distribution of migrants between potential destinations. About half of the world's international migrants go to the United States, where they account for 40% or more of its annual population growth. In the late 1990s foreign-born workers made up half of the net growth of the United States' labour force (Kent and Haub 2005: 10 and 17).

The effects of migration on population aging inevitably vary according to the shape of each country's age structure, but demographic models again illustrate the range of possibilities. Figure 2.6 shows the projected effects of migration, after 50 years, on three contrasting model populations spanning the current range in total fertility rates in developed countries from 2.1 (100% replacement) to 1.6 (75%) and 1.0 (50%). As in Fig. 2.5, the three model populations assume a constant life expectancy of 80 years for females and 77 years for males.

Rates of migration depend on its social and political acceptability, the availability of adequate numbers of suitable migrants and the ability of the country to accommodate and employ them. The net migration rate of 5 per 1,000 in Fig. 2.6 approximates peak figures for migration to the United States in the early twenty-first century. However, it is unlikely that many countries could sustain for long a net migration rate of 5 per 1,000. Figure 2.6 therefore includes the effect of an intermediate annual net migration rate of 2.5 per 1,000, as well as zero net migration. The age structure of net migration is held constant at that employed in 'middle-range' projections for the United States, and migrant fertility is constant at the level for the destination, because migrant fertility commonly converges to that of the receiving country. Higher migrant fertility would boost growth arising from migration, potentially provoking concerns about migrant communities expanding their overall representation in the population.

The volume of net migration in each model remains constant through time because it is based on the initial population. This is useful for illustrative purposes as it omits complications arising from interactions between population size and the number of migrants. Alternatively, the migration rate could have been applied to the population at each 5 year step in the projection, resulting in an increasing volume of migration through time in growing populations and a decreasing volume in declining populations. The assumption of a constant volume of migration is more neutral because it neither assumes increasing demand for immigrants and expanding capacity to absorb them in growing populations, nor the opposite in declining populations – where migrants may be most needed.

For the three levels of fertility and the three levels of migration, Fig. 2.6a shows projections, after 50 years, of the percentages aged 65 and over. The main finding from Fig. 2.6a is that when fertility is below replacement, migration can have an appreciable rejuvenating effect on the age structure: higher rates of migration lead to lower percentages in the older ages. Also, the lower the fertility rate, the greater migration's rejuvenating effect. For example, in a population with fertility at half the replacement level, the percentages aged 65 and over would rise to 37% in the absence of migration, but to only 26% after 50 years of high migration, were it feasible. The rejuvenation would be more pronounced if migrant fertility was above that of the country of settlement.

However, the further the fertility falls below replacement, the less effective even high migration is in averting population decline. This is illustrated in Fig. 2.6b, which shows the size of the total population after 50 years, relative to the initial population. Thus in the population with fertility at 75% of replacement, numbers are 94% of the initial population size after 50 years of high migration, compared with 59% for the population with fertility at 50% of the replacement level.

In Fig. 2.6c the age-sex structures, for ages 0–4 to 85 and over, summarize the effects of high migration on the three model populations. Figure 2.6c depicts the initial and final (50 years later) age structures of populations with each level of fertility, assuming a high migration rate and an initial population of one million. Migration at a rate of 5 per 1,000 augments the size of the population with 100% replacement fertility, but does not change the shape of its age profile greatly, apart from increasing the representation of young adults – the main migrant group. In the population with 75% replacement, high migration maintains much of the original shape as well as the size of the population. This confirms that, other things being equal, high migration can be an effective process of demographic replacement in populations with long-run total fertility rates at or above 1.6 children per woman. Raising the migration rate to just over 6 per 1,000 would keep the model's total population numbers constant at around one million. In contrast, the age structure with high migration and very low fertility (50% of replacement) confirms that migration cannot avert a major decline in numbers in countries with very low fertility, although it does restore a degree of balance to a top-heavy age structure.

In light of these findings, a continuation of very low fertility in some European and Asian countries will place them beyond assistance from migration in circumventing substantial population decline. However, they could benefit from its effects in reducing peak percentages in the older ages. While the models are indicative both of circumstances to avoid and of beneficial policy strategies, they can do no more than illustrate the long run effects of holding demographic rates constant. In real populations processes of change are seldom constant for long and countries do not have initial age structures exactly matching the models. Nevertheless, national population projections along the lines of those in Fig. 2.6 are important aids to decision-making through revealing consequences of particular trends.

2.6 The Tempo of Aging

Statistics on the time taken for the percentage aged 65 and over to double from 7 to 14 – reveal the protracted nature of population aging in the past. The interval for France was 115 years (1865–1980), for Sweden 85 years (1890–1975), the United States 69 years (1944–2013) and the United Kingdom 45 years (1930–1975) (Kinsella and Phillips 2005: 13). A striking development, foreshadowed in the United Nations medium variant projections, is the increasing tempo of aging. The pace of change is likely to quicken, partly because the large generations, born in baby booms and the population explosion, are growing older, and partly because of the extent of fertility decline in some countries. Associated with the latter is the undermining of age structures and the strengthening of negative momentum. Their continuation in coming decades would produce, around mid-century in some societies, exceptionally rapid population aging and unprecedented strains on welfare systems.

This is shown by further statistics on the tempo of aging, measuring the time taken to add five to the percentage in the older ages. The figures were calculated from United Nations (2001) estimates and projections from 1950 to 2050. Typically, the shift from 10% to 15% is protracted – the figure is 45 years for the United States, and 33 years for more developed regions overall. The next step, from 15% to 20% is expected to be generally more rapid – 18 years for developed regions – although Germany, the United Kingdom and several other countries may have intervals roughly double this figure. The shift from 20% to 25% may take only between 10 and 16 years in the world's oldest countries. An exception is Japan, which has the shortest intervals of 8–12 years through the initial levels of aging because of its early experience of low fertility. Japan, therefore, had the most rapid rate of aging from about the early 1980s to the early 2010s. After this, the projections indicate a higher tempo of aging in Italy and Spain: in intervals of 10 years or less they might complete the steps from 25% to 30% and from 30% to 35%. The speed of aging here highlights the prospect of sudden and extreme consequences awaiting countries with persistent very low fertility. The highest figures are the most speculative, however, as they refer to developments in the 2040s. Other projections imply a somewhat lower tempo around mid century, that is if fertility in developed countries is a little above that projected in 2000. A key point is that if hyper-aging is allowed to develop, shifts to higher levels of aging are likely occur at progressively shorter intervals. The challenges of adjusting to such circumstances would far exceed current experience.

2.7 Conclusion

Unbalanced age structures, with inherent potential for excessive aging and rapid population decline, are products of the new demography. There is now greater uncertainty about the future than seemed likely when the classical demographic transition was the dominant model, because the potential range of variation is wider and the need for varied policy interventions is greater as well. What may seem minor features today can expand into major concerns. Long range projections raise issues that are not simply matters for future generations to address, but are essential input to present-day planning. Some dismiss long range projections as science fiction, especially if they include the unborn children of unborn parents. This overlooks the experimental applications of projections in clarifying the long-range implications of different population trajectories. They assist in identifying the advantages and disadvantages of current trends in light of possible developments at a number of points into the future. Also overlooked is the fact that certain changes, such as growth in the numbers of middle aged and older people, are built into a population's age structure and will be relevant to policy making for many decades ahead. Thus projections are widely recognized and carefully scrutinized aids to anticipating developments that flow from the present situation and which, without change or redirection, may have the potential to become socially disruptive. Looking to the long-term future is all the more relevant because it seems unlikely that countries with the lowest fertility will return to levels near replacement by 2025. Rising life expectancy, a widely desired goal, is similarly fostering the emergence of societies in which, for the first time in history, older people will outnumber children.

References

Barclay, G. W. (1966). *Techniques of population analysis*. New York: Wiley.

Bongaarts, J., & Bulatao, R. A. (Eds.). (2000). *Beyond six billion: Forecasting the world's population*. Washington, DC: National Academy Press.

Caldwell, J. C. (2003). The contemporary population challenge. *Population Bulletin of the United Nations, Completing the Fertility Transition*, Special Issues, Nos 48/49, 81–88.

Castles, F. G. (2004). *The future of the welfare state: Crisis myths and crisis realities*. Oxford: Oxford University Press.

Castles, S., & Miller, M. J. (2003). *The age of migration: International population movements in the modern world*. New York: Guildford Press.

Coale, A. J., & Demeny, P. (1983). *Regional model life tables and stable populations*. New York: Academic Press.

Cowgill, D. O. (1970). The demography of ageing. In A. M. Hoffman (Ed.), *The daily needs and interests of older people* (pp. 27–69). Springfield: Charles C. Thomas.

Kaneko, R. (2008). Statistical foundations of population projections. In F. Coulmas, H. Conrad, A. Schad-Seifert, & G. Vogt (Eds.), *The demographic challenge: A handbook about Japan* (pp. 41–59). Leiden/Boston: Brill.

Kent, M. M., & Haub, C. (2005). Global demographic divide. *Population Bulletin, 60*(4).

Keyfitz, N. (1971). On the momentum of population growth. *Demography, 8*, 71–80.

Kim, Y. J., & Schoen, R. (1997). Population momentum expresses population aging. *Demography, 34*(3), 421–427.

Kinsella, K., & Phillips, D. R. (2005). Global aging: The challenge of success. *Population Bulletin, 60*(1).

Kirkwood, T. (2001). Lecture 1: The brave old world, the end of age. *Reith lectures 2001*. London: BBC. http://www.bbc.co.uk/radio4/reith2001/lecture1.shtml. Accessed April 2009.

Myers, G. C. (1990). Demography of aging. In R. H. Binstock & L. K. George (Eds.), *Handbook of aging and the social sciences* (3rd ed., pp. 19–44). San Diego: Academic.

Notestein, F. (1954). Some demographic aspects of aging. *Proceedings of the American Philosophical Society, 98*, 38–46.

Population Reference Bureau. (2010). *World population data sheet 2010*. Washington, DC: Population Reference Bureau.

Potter, R. G., Wolowyna, O., & Kulkarni, P. M. (1977). Population momentum: A wider definition. *Population Studies, 31*, 555–569.

Rogers, A. (1989). The elderly mobility transition: Growth, concentration and tempo. *Research on Aging, 11*(1), 3–32.

Rowland, D. T. (1996). Population momentum as a measure of ageing. *European Journal of Population, 12*, 41–61.

Rowland, D. T. (2003). *Demographic methods and concepts*. Oxford: Oxford University Press.

United Nations. (2000). *Replacement migration: Is it a solution to declining and ageing populations?* New York: Population Division, Department of Economic and Social Affairs, United Nations Secretariat, ESA/P/WP.160.

United Nations. (2001). *World population prospects: The 2000 revision (CD, disk 1: standard set)*. New York: United Nations, Department of Economic and Social Affairs, Population Division.

United Nations. (2009a). *Assumptions underlying the results of the 2008 revision of world population prospects*. New York: United Nations. http://esa.un.org/UNPP/index.asp?panel=4. Accessed April 2010.

United Nations. (2009b). *World population prospects: The 2008 revision population database*. New York: United Nations, Department of Economic and Social Affairs, Population Division. http://esa.un.org/UNPP/index.asp?panel=2. Accessed April 2010.

Chapter 3
National Trends

In the past 25 years, the number of people of pensionable age (65 and over) in OECD countries rose by 45 million, but the population of working age rose by 120 million. As a result, population ageing has so far posed no major economic or social problems for our societies. This will change dramatically in the next 25 years when the number of persons of pensionable age will rise by a further 70 million, while the working-age population will rise by only five million.

(OECD 1998: 1)

3.1 Population Profiles

As shown in Chap. 2, demographic trends and their implications change dramatically when societies experience very low birth rates and extended longevity. This 'new demography of aging' is reshaping national age structures. According to classical demographic transition theory, national age structures gradually evolve from an initial triangle or pyramid shape to a final pentagon with vertical sides, as the representation of all but the oldest age groups evens up. However, in contemporary developed countries, new types of national age structures are gradually emerging, notably some variously described as having the profile of an 'aircraft carrier', an 'overhanging cliff', an 'upside down pyramid' or a 'coffin'. These top-heavy, narrow-based forms – indicative of marked imbalances in societies – will become widespread unless current trends are reversed. Changes in age structures immediately underlie much of the transformation that aging societies are experiencing. The patterns of age structure change reveal the processes at work in different countries, enable comparisons of trends through time, and permit generalizations about the past and future of population aging. This chapter first discusses trends in national and regional age structures at four points in time, using United Nations estimates for 1950, 1975, and 2000 together with medium variant projections for 2025. These four 'slices' provide a manageable summary of developments in each area over a 75 year period. They reflect the

D.T. Rowland, *Population Aging: The Transformation of Societies*, International Perspectives on Aging 3, DOI 10.1007/978-94-007-4050-1_3,

cumulative effects of past changes and features that will influence future experience. Projections for 2050 are not included in this more detailed analysis because they are much more speculative than those for 2025. Later sections examine consequences of age structure transformation evident in labour force changes and generational shifts affecting the ability of societies to maintain their populations.

3.2 Types of Age Structures

The analysis of changes in national age structures is based on a classification of age distributions, at the above four points in time, for 33 United Nations regions and sub regions and 139 countries with populations of a million or more in the year 2000. The classification, which grouped similar age structures, was obtained using the statistical technique of cluster analysis. Figure 3.1 presents the results of the classification in terms of groups of age structures associated with (i) above replacement fertility, (ii) below replacement fertility, and (iii) discontinuities in age structure change, such as baby booms. The chart presents an average age structure for each group, as well as statistics on median ages, the percentages 65 and over and the number of cases in each group.

The age structures in the first row of Fig. 3.1 correspond to the sequence of changes predicted in the demographic transition; they are mainly from developing countries. The transition has a theoretical end point in the post-transition stage, where a new balance between birth and death rates produces zero growth, although no country has reached this stage. The model population on the right of the first row of Fig. 3.1 represents a post-transition age structure: it has replacement level fertility and zero growth. It is also a long-lived population, with life expectancies of 85 for females and 79 for males, which is consistent with current prospects – compared with 75 for females in the original, classical transition. Its median age of 41 years is less than that for most of the age structures associated with below-replacement fertility. The shape of a post-transition age structure, if it existed, would vary according to the life expectancy reached at the end of the transition: the higher the life expectancy, the higher the ultimate percentage in the older ages. The post-transition age structure represents what is, demographically, most sustainable in the long-term – because it avoids problems of growth and decline, maintains a balance between the numbers in different generations, and avoids excessive aging. Its profile is superimposed on the others in Fig. 3.1 to provide a benchmark for comparisons. The model draws attention to deficits and surpluses in particular age structures. Compared with the model population, the types of age structures associated with above-replacement fertility have relatively high proportions in the young ages, a situation currently forewarning of substantial future rises in the numbers of older people.

In contrast, the age structures in the middle row reflect that below replacement fertility creates deficits in the child ages and raises the percentage of older people. Deficits can flow on to labour force and family formation ages, inducing a cycle of self-perpetuating decline. Whereas the first demographic transition forecast that national age structures would converge to a rectangular form, the second demographic

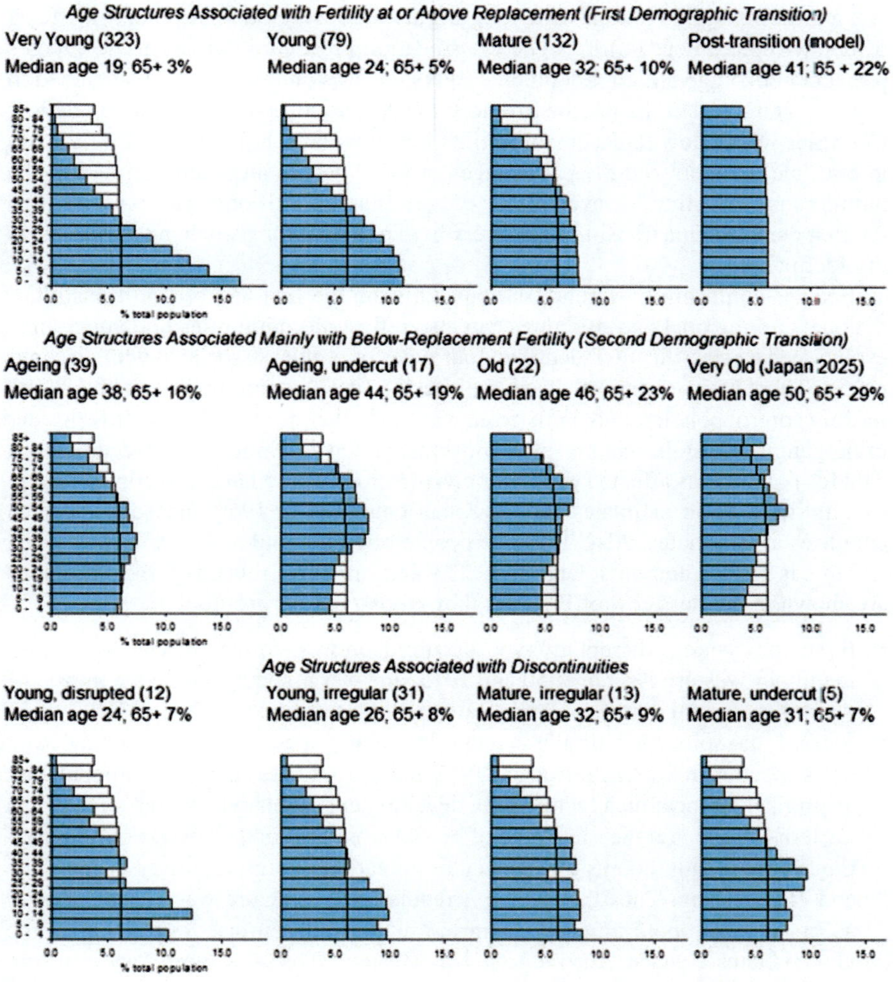

Fig. 3.1 Main types of national and regional age structures, 1950, 1975, 2000 and 2025 (*Note*: The figures in brackets show the number of age structures of each type. There were 172 countries and regions in the classification, giving a total of 688 age structures; 15 age structures were ungrouped or belonged to groups of less than 5) (Source: Cluster analysis of data from United Nations 2001)

transition foreshadows diversity among the oldest populations, depending on the extent to which their age profiles are undercut. There is no specific end point to these changes because the situation is inherently unstable: the potential for change ranges from moderate rejuvenation to extreme decline.

An emerging prospect for many developed countries is a long-term struggle to maintain demographic viability. Most regions of the Western world had mature or aging profiles in 2000 and are projected to have mainly old profiles in 2025. The oldest population in the data set, Japan in 2025, has a median age of 50 years, and a tapering age pyramid, which highlights the severe population aging effects of below-replacement fertility. The classification procedure did not group Japan's projected 2025 age structure with any others – it was unique – but it is included in Fig. 3.1, at the end of the middle row, to show the structure of the oldest population. Countries with below replacement fertility currently have high potential for growth in both the numbers and the percentages of older people although, ultimately, the numbers may decline. Conversely, countries that are still only part way through the first demographic transition have very high potential for growth in their numbers of older people.

Besides conforming to trends anticipated from the first and second transitions, national and regional age structures can also reflect discontinuities in demographic evolution. The classification identified four types of profiles associated with discontinuities. These have arisen partly from the effects of baby booms, migration booms and fertility control policies, and partly from wars and other disasters leading to episodes of heightened mortality and refugee movements. Uncertainties and inaccuracies in data for populations affected by war or civil disorder would have also contributed to irregularities, as in estimates for European countries in 1950 and some African countries at later dates. Also, illiterate people often do not know how old they are, and this is more common at later ages. The age structures reflecting discontinuities are shown in the third row of Fig. 3.1; they belong to four groups:

- First, the 'young, disrupted' type occurred only in former Soviet Republics, mainly in Western Asia in 1950 and 1975, and was associated with the impact of war losses, social disruption and relatively high fertility.
- Second, the somewhat similar 'young, irregular' type characterized many other age structures in Eastern Europe in 1950 and 1975. Losses in the adult population again made truncation a factor in the development of this type of age structure in Eastern Europe. Yet the same type of profile arose also in the context of rejuvenation through baby booms due especially to earlier and more universal marriage and childbearing. Thus the 'young, irregular' age structure was conspicuous in 1975 in baby boom-affected countries in Western Europe together with the United States, Canada, Australia and New Zealand. The same type of age structure occurred as well in Israel in 1975 and 2000 as a consequence of immigration and high birth rates. All age structures of this type were weighted to the younger ages, although not necessarily with a single pronounced step below 30–34.
- Third, the 'mature irregular' age structures occurred in 1950 in war affected countries of Europe, including France, Belgium, Germany and Austria, as well as in parts of Eastern Europe in 1950 and 1975.
- Finally, the 'mature undercut' profile appeared only in the year 2000 and arose from accelerated fertility decline. It was characteristic in the United Nations region of East Asia, notably China and South Korea.

Inevitably, the classification does not show the full extent of diversity because it collapsed the whole range of structural variations into groups of profiles that most resembled each other. This means that the average age structure for each group is not identical to particular national age structures within the group. Diversity within groups

Table 3.1 Summary of age structure changes in United Nations regions, 1950–2025

United Nations region	1950	1975	2000	2025
World	Very young	Very young	Young	Mature
More developed regions	Mature	Mature	Aging I	Old I
Less developed regions	Very young	Very young	Young	Mature
Least developed regions	Very young	Very young	Very young	Very young
Africa	Very young	Very young	Very young	Very young
Asia	Very young	Very young	Young	Mature
Eastern Asia	Very young	Very young	Mature, undercut	Aging I
Latin America & the Caribbean	Very young	Very young	Young	Mature
Europe	Mature	Mature	Aging I	Old I
Eastern Europe	Young, irregular	Mature, irregular	Mature	Aging, undercut
Northern Europe	Mature	Mature	Aging I	Old I
Southern Europe	Mature	Mature	Aging II	Old II
Western Europe	Mature, irregular	Mature	Aging II	Old II
Northern America	Mature	Young, irregular	Aging I	Aging II
Oceania	Mature	Young, irregular	Mature	Aging I

Source: Cluster analysis of data from United Nations (2001)
Note: The 'aging' and 'old' groups were split where the percentages aged 65 and over were at or below the average for the group (aging I, old I), and above the average (aging II, old II)

was partly addressed by providing more detail, in Tables 3.1 and 3.2, for two of the advanced stages of the population aging process. Thus the 'aging' and 'old' groups were split according to whether each population's percentages 65 and over were at or below the mean for the group, (aging I, old I) or above the mean (aging II, old II). This highlights, for example, the relatively high levels of aging already developing in Southern and Western Europe, as well as in their constituent countries. Because assumptions about fertility are usually the main cause of differences between sets of national projections, the 2025 classifications would vary somewhat if a different series of projections was used. For example, comparing projections for Europe from the 2008 revision of the UN projections with those for the 2000 revision used here, the higher fertility assumptions in 2008 raise the proportion in the child age groups slightly (up 1.6% in 2025) and decrease the older age groups (down 1.9% in 2025). These changes in the apparent extent of aging could alter again in further revisions of the projections.

3.3 Changes Through Time

Tables 3.1 and 3.2 show the results of the classification for regions and selected countries respectively. The latter refers to the subset of 38 countries, described in the Preface, with the oldest populations or the largest aged populations, together

Table 3.2 Age structure changes in selected countries, 1950–2025

Countries	1950	1975	2000	2025
North America and Australasia				
United States	Mature	Young, irregular	Aging I	Aging II
Canada	Mature	Young, irregular	Aging I	Old I
Australia	Mature	Young, irregular	Aging I	Aging II
New Zealand	Mature	Young, irregular	Aging I	Aging II
Northern and Western Europe				
United Kingdom	Mature	Mature	Aging II	Old I
France	Mature, irregular	Mature	Aging I	Old I
Switzerland	Mature	Mature	Aging I	Old II
Belgium	Mature, irregular	Mature	Aging II	Old II
Netherlands	Mature	Young, irregular	Aging I	Old I
Denmark	Mature	Mature	Aging I	Old I
Norway	Mature	Mature	Aging I	Old I
Sweden	Mature	Mature	Aging II	Old II
Finland	Mature	Young, irregular	Aging I	Old II
Germany	Mature, irregular	Mature	Aging II	Old II
Austria	Mature, irregular	Mature	Aging I	Old II
Southern Europe				
Italy	Mature	Mature	Aging II	Old II
Greece	Young, irregular	Mature	Aging II	Old II
Slovenia	Mature, irregular	Young, irregular	Mature	Aging, undercut
Spain	Mature	Mature	Aging II	Old II
Portugal	Mature	Mature	Aging I	Old I
Eastern Europe				
Poland	Young, irregular	Young, irregular	Mature	Aging, undercut
Czech Republic	Mature, irregular	Mature	Mature	Aging, undercut
Croatia	Young, irregular	Mature	Mature	Old I
Bulgaria	Young, irregular	Mature	Mature	Aging, undercut
Russian Fed.	Young, irregular	Mature, irregular	Mature	Aging, undercut
Ukraine	Young, irregular	Mature, irregular	Mature	Aging, undercut
Hungary	Mature, irregular	Mature	Mature	Aging, undercut
Latvia	Mature, irregular	Mature	Mature	Aging, undercut
Asia and Latin America				
Japan	Very young	Mature	Aging II	(very old)
Singapore	Young	Young	(mature, v. undercut)	(old, irregular)
Hong Kong	(v. young, irregular)	Young, irregular	(mature, v. undercut)	Aging, undercut
China	Very young	Very young	Mature, undercut	Aging I
Indonesia	Very young	Very young	Young	Mature
India	Very young	Very young	Young	Mature
Bangladesh	Very young	Very young	Very young	Mature
Pakistan	Very young	Very young	Very young	Young
Mexico	Very young	Very young	Young	Mature
Brazil	Very young	Very young	Young	Mature

Source: Cluster analysis of age data (standardized) from United Nations (2001)
Note: The 'aging' and 'old' groups were divided into populations with percentages aged 65 and over that were at or below the average for the group (aging I, old I), or above the average (aging II, old II). Bracketed descriptions denote ages structures occurring only once or twice in the classification

with a further four countries of particular interest for comparisons. The countries are listed mainly by region in Table 3.2, for comparability with later analyses, especially the grouping of countries in Chap. 17. Dominating the global pattern are trends in developing countries, which now have by far the greatest numbers of older people. Hence the age structure classifications for the less developed regions are the same as for the world as a whole, with very young profiles giving way to young by 2000 and mature by 2025. The same pattern occurred in the United Nations regions of Asia, except for Eastern Asia, as well as in all the constituent regions of Latin America and the Caribbean. This was also true for 4 out of the 7 countries with the largest projected aged populations in 2025, namely, India, Indonesia, Mexico and Brazil.

Nevertheless, differences in the pace of fertility change are producing variations among developing countries. The least developed regions, including much of sub-Saharan Africa, remain very young over the whole 75 years. In contrast, China, Hong Kong and Singapore are aging relatively rapidly (Table 3.2). Although these three have high growth economies, the UN includes them in the less developed regions. After China, Japan – with a population in 2010 of 127 million – is the next most populous country in Eastern Asia. Japan is ahead of China in terms of aging, because of its higher life expectancy and because its rapid fertility decline after the Second World War brought an early start to its experience of below replacement fertility. Out of all of Asia, Latin America and Africa, regional population aging is pronounced only in Eastern Asia.

Meanwhile, the more developed regions had mature profiles in 1950 and 1975, leading to aging and old profiles in 2000 and 2025 respectively. Their mature age structures in 1950 and 1975 correspond to those projected for much of Asia and Latin America in 2025. The recent history of more developed regions reveals heightened aging associated with the crossover from the first to the second demographic transition. When populations with mature age structures experience sustained below-replacement fertility, undercut and rapidly aging profiles soon emerge.

Europe's age-structure changes are consistent with the more developed regions pattern overall. In coming decades Europe is likely to have some of the world's highest percentages in older ages. Its age structure is also developing greater inherent momentum of decline and all regions of Europe are aging significantly. Whereas Northern Europe's pattern of age structure change is the same as that for the sub-continent overall, Western and Southern Europe both have somewhat older age structures in 2000 and 2025. Because of the size of their populations, Germany, Italy and Spain have a considerable impact on the aging of these two regions. Countries with the youngest age structures in Europe in 2025 are the United Kingdom, France, the Netherlands, Denmark and Norway.

Like Eastern Asia, Eastern Europe has had a distinctive experience of age structure change and aging (Table 3.1). Its young irregular and mature irregular profiles in 1950 and 1975 were the net outcome of turbulent national histories, including the effects of the Second World War, subsequent refugee movements and continuing political oppression and upheaval. Peaks and valleys in Eastern Europe's age structures have generated sharp changes in the numbers advancing from one age group to the next. Despite these circumstances, fertility in Eastern Europe remained relatively high over this period (TFR 2.9 to 2.2) and even in 2000 the region's age structure was still mature. The shift to an aging undercut age structure by 2025, is indicative of an expected acceleration of aging associated with low fertility bringing a falling representation of children and people in the working ages. Most of the individual Eastern European countries in Table 3.2 also have this type of age structure in 2025.

More moderate patterns of aging are evident in the United States, Australia and New Zealand which had prolonged baby booms after the Second World War and whose fertility has been projected to remain relatively high for developed countries. Thus despite having mature age profiles in 1950, they underwent rejuvenation leading to young irregular profiles in 1975. Similarly in 2025 their age structures are projected to be younger than that for developed regions overall (Table 3.2).

3.4 Labour Force Changes

The sustainability of higher levels of population aging will depend substantially on the size and productivity of national labour forces, which low birth rates and longer life have the potential to undermine and overburden. Prospective changes in the numbers of working age are indicators of the ability of societies to support the financial demands of aging populations. Total numbers aged 15–64 are an initial measure of labour resources despite the fact that many are engaged in domestic duties or full-time education, or are unemployed or unable to work. In the future, higher labour force participation among adults of all ages may augment labour supplies independently of the size of the 'working age' population.

United Nations projections highlight the impact that low birth rates may have on labour force numbers, especially in the longer term (Table 3.3). At a regional level, Europe is the only major region projected to experience a decline in the working

Table 3.3 Changes in total and working age populations, United Nations regions and selected countries, 2000–2050

	Total population		Working ages (15–64)	
Region/country	Numbers 2000 (thousands)	% change 2000–2050	% change 2000–2025	% change 2025–2050
Global summary				
World	6,115,367	49.6	36.9	11.5
More developed regions	1,194,970	6.7	0.5	−7.9
Less developed regions	4,920,401	60.0	46.5	15.0
Africa	819,461	143.9	88.0	56.5
Asia	3,698,294	41.5	37.3	5.0
Eastern Asia	1,472,443	8.7	13.1	−14.0
Europe	726,570	−4.9	−5.1	−14.7
Eastern Europe	304,085	−21.1	−13.5	−21.4
Northern Europe	94,357	19.3	7.6	1.9
Southern Europe	145,119	5.9	2.9	−17.0
Western Europe	183,002	1.0	−3.5	−11.7

(continued)

Table 3.3 (continued)

Region/country	Total population: Numbers 2000 (thousands)	Total population: % change 2000–2050	Working ages (15–64): % change 2000–2025	Working ages (15–64): % change 2025–2050
Latin America and the Caribbean	521,228	39.9	39.0	2.3
Northern America	318,657	40.7	19.2	8.9
Oceania	31,161	64.7	34.3	18.8
North America and Australasia				
United States	287,842	40.3	19.4	9.1
Canada	30,686	44.7	17.0	6.6
Australia	19,170	49.8	21.6	9.9
New Zealand	3,869	38.3	20.8	5.3
Northern and Western Europe				
United Kingdom	58,909	22.8	9.8	4.2
France	59,129	14.4	3.7	−3.5
Switzerland	7,185	18.5	5.0	−2.2
Belgium	10,193	12.7	2.5	−3.9
Netherlands	15,914	9.3	0.0	−5.6
Denmark	5,336	4.0	−2.1	−4.3
Norway	4,484	32.6	16.7	4.9
Sweden	8,860	19.3	6.2	3.7
Finland	5,173	5.2	−4.7	−3.8
Germany	82,073	−14.1	−11.3	−21.8
Austria	8,006	6.4	1.9	−12.7
Southern Europe				
Italy	57,114	−0.1	−1.7	−19.5
Greece	10,944	0.0	−2.8	−17.0
Slovenia	1,987	−1.6	−6.8	−16.3
Spain	40,267	27.3	15.7	−14.1
Portugal	10,227	−2.1	−0.4	−20.7
Eastern Europe				
Poland	38,431	−16.7	−8.9	−23.5
Czech Republic	10,224	0.7	−4.7	−13.6
Croatia	4,504	−15.1	−11.4	−18.4
Bulgaria	8,009	−32.7	−20.3	−31.0
Russian Federation	146,672	−20.8	−14.2	−19.8
Ukraine	48,870	−28.3	−18.8	−23.6
Hungary	10,214	−12.5	−10.0	−15.7
Latvia	2,371	−21.9	−14.7	−19.6
Asia				
China	1,266,954	11.8	16.5	−12.7
India	1,042,592	54.8	53.6	12.9
Japan	126,703	−19.8	−17.0	−27.7
Singapore	4,017	30.0	21.5	−15.3
Hong Kong	6,668	29.3	9.3	−7.9

Source: Calculated from United Nations (2009)

ages 15–64, namely 5% 2000–2025 and 15% 2025–2050. At a sub-regional scale only Northern Europe, which includes the UK and the Nordic countries, maintains labour force growth until mid-century. Heavy sustained losses occur in Eastern Europe as net outward migration reinforces depopulation. Labour force decline and overall population decline often go together in Eastern Europe with Bulgaria, the Russian Federation and Ukraine having the highest losses (Table 3.3). Southern and Western Europe may also have high labour force losses in the second quarter of the century due to the cumulative effects of low fertility. Germany is the only Western country with substantial overall population decline and labour force decline projected for the first half of this century.

Perhaps surprisingly, Italy's projected total population at mid-century is close to its total at the start of the century, but the working ages decline by 20% in the second quarter (Table 3.3). The size of Italy's population at 2050 is reasonably consistent with 2007 projections from the Italian National Institute of Statistics (ISTAT 2007). However the figures differ substantially from a previously projected 25% decline in its total population 2000–2050 (United Nations 2001). The absence of overall population decline in the newer figures is due to higher projected total fertility rates of 1.4–1.7 for 2010–2050, instead of 1.2–1.6, together with life expectancy rising to 88.4 for females in 2050 instead of 85.6. These changes maintain moderate overall growth until about 2015, after which numbers decline continuously to 2050. Net immigration of 150,000 per annum after 2015, which is half the figure in the 2000 series, becomes increasingly inadequate to offset the depopulating effects of Italy's low fertility regime in which deaths are expected to exceed births by a greater margin each year (ISTAT 2007). Thus the similarity between Italy's population totals at 2000 and 2050 masks a situation where there is no natural increase, a top-heavy age structure is developing as growth occurs only at the apex of the age pyramid, and there is a considerable decline in the numbers of working age. Italy's level of aging is high at 33% aged 65 and over in 2050. Greece, Portugal, Spain and Slovenia are other Southern European countries with substantial labour force declines after 2025 and levels of aging above 30% by 2050 (Table 3.3).

Previously (United Nations 2001) the same was anticipated in some other Western and Northern European countries, but higher fertility and life expectancy assumptions have greatly moderated this, as well as reducing projected levels of aging at 2050. Nevertheless, in Europe only the United Kingdom, Norway, and Sweden have appreciable labour force growth in the 2008 projections. Much greater population and labour force growth is anticipated for Australia, New Zealand, Canada and the United States. Yet even these countries are confronting economic concerns arising from population aging, which emphasizes the challenge that awaits countries less favourably placed in terms of maintaining their labour forces. Among the developed countries, Japan stands out as having the oldest population, as well as the greatest declines in total numbers and people of working age outside of Eastern Europe (Table 3.3).

Meanwhile, great increases in the numbers of people of labour force age are projected for Africa, Asia and Latin America. Small island states of the Caribbean and Oceania will also experience relatively high increases that will be difficult to absorb locally.

Most dramatic is the expectation that India's population of working age (15–64 years) could increase by about 54% in the first quarter of this century. India is overtaking China as the world's most populous country. China's labour force may grow by 16% (141 million) 2000–2025, but, by 2050, the numbers are likely to fall close to the 2000 total because of the smaller size of cohorts born since the 1970s.

Migration from developing to developed countries offers some mutual benefits, including migrants' remittances to relatives at home. However, the number of migrants that countries with the oldest populations can absorb, while maintaining economic and social stability, could be too low to offset their own labour force deficits, or to have any appreciable effect on numbers in the most populous countries of emigration. At the same time, the potential supply of immigrants is much more limited than population statistics imply, because the workers now in greatest demand by countries of immigration – people with occupational skills and qualifications – are in shortest supply and are often those that the places of origin can least afford to lose.

3.5 Generational Changes

Generational changes, entailing shifts in the relative size of cohorts at different stages of life, have further implications for the ability of societies to maintain themselves. The projections suggest there will be dramatic variations through time in indicators of this, such as national labour market entry-exit ratios. These reveal generational differences underlying labour force decline as the numbers at labour force entry age (15–24) fall below the numbers at labour force exit age (55–64). The regional figures in Table 3.4 point to an almost world-wide shift in this direction, with only Africa and parts of other less developed regions defying the trend. Unprecedented labour shortages could occur in many countries as aging peaks, necessitating higher and longer labour force participation as well as higher labour productivity.

In 2000, only Germany, Italy and Japan had fewer at the entry ages than at the exit ages. The ratios were favourable for labour force growth in all the other countries. Some had particularly high ratios in 2000, including Hong Kong and Singapore and parts of Eastern and Southern Europe (Table 3.4). By 2025 the situation could change dramatically in the developed countries, with only the United States, Australia and New Zealand projected to have either an excess of entries or a balance of entries and exits. Germany, Austria and Japan, together with Hong Kong, Singapore and several Southern European countries, might all have ratios of less than 70 entries per 100 exits. These reversals mainly reflect the impact of deficits in generation building. The situation in 2050 is far more uncertain but the projected upswing in total fertility rates – for example to between 1.7 and 1.9 in the sub-regions of Europe in 2050 – would ameliorate the situation. This reflects that age structure changes are sensitive to seemingly minor alterations in birth rates, as even a positive change of 0.2 in a total fertility rate entails an appreciable number of extra births

Table 3.4 Indices of generational changes, United Nations regions and selected countries, 2000, 2025 and 2050

	Labour market entry-exit ratio[a]			Aging index[b]			Aged dependency ratio[c]		
Region/country	2000	2025	2050	2000	2025	2050	2000	2025	2050
Global summary									
World	275	157	113	22	43	83	11	16	25
More developed regions	134	85	84	78	131	170	21	33	45
Less developed regions	338	177	119	15	33	72	8	13	23
Africa	489	399	225	8	12	26	6	7	11
Asia	297	144	96	19	44	97	9	15	27
Eastern Asia	207	82	71	31	85	164	11	22	40
Europe	132	78	77	84	137	182	22	32	47
Eastern Europe	157	84	67	71	120	167	19	28	43
Northern Europe	118	89	92	82	115	144	24	31	39
Southern Europe	125	70	78	105	157	222	25	34	57
Western Europe	106	72	84	93	160	197	24	38	51
Latin America and the Caribbean	343	159	96	18	48	114	9	16	31
Northern America	161	106	97	58	100	130	19	29	36
Oceania	194	135	120	38	66	98	15	23	30
North America and Australasia									
United States	162	109	99	57	97	126	19	29	35
Canada	147	78	85	66	127	164	18	32	43
Australia	149	98	98	61	107	143	19	30	40
New Zealand	157	99	92	52	97	140	18	29	39
Northern and Western Europe									
United Kingdom	116	89	95	84	112	139	24	31	38
France	140	97	99	86	135	166	25	37	47
Switzerland	106	72	91	89	148	166	23	35	45
Belgium	118	81	94	97	134	166	26	36	46
Netherlands	118	81	98	73	135	162	20	35	44
Denmark	100	87	100	80	130	148	22	34	40
Norway	128	92	94	75	112	146	23	31	40
Sweden	102	90	88	93	126	146	27	36	41
Finland	119	86	89	82	145	164	22	40	44
Germany	85	54	67	105	203	258	24	40	59
Austria	106	64	76	91	158	210	23	34	52
Southern Europe									
Italy	99	64	77	129	193	247	27	39	62
Greece	132	68	75	108	169	230	24	35	57
Slovenia	140	67	72	88	158	212	20	35	54
Spain	145	72	85	114	137	215	25	32	59
Portugal	134	72	75	100	169	242	24	35	59
Eastern Europe									
Poland	196	82	57	64	150	235	18	32	52
Czech Republic	143	82	73	84	135	179	20	32	48
Croatia	123	73	73	92	152	195	23	35	49
Bulgaria	123	71	66	106	155	212	24	34	55

(continued)

Table 3.4 (continued)

Region/country	Labour market entry-exit ratio[a]			Aging index[b]			Aged dependency ratio[c]		
	2000	2025	2050	2000	2025	2050	2000	2025	2050
Russian Federation	162	88	71	68	109	145	18	27	39
Ukraine	128	77	68	80	116	154	20	29	42
Hungary	131	86	73	90	138	177	22	31	44
Latvia	117	75	62	86	121	168	23	30	44
Asia									
China	229	84	71	26	74	153	10	19	38
India	361	207	102	12	30	75	7	11	20
Japan	98	68	66	118	269	337	25	50	74
Singapore	168	52	61	33	185	291	10	35	58
Hong Kong	189	45	57	65	186	289	15	34	58

Source: Calculated from United Nations (2009)
[a]Labour market entry-exit ratio: persons 15–24 per hundred 55–64
[b]Aging index: aged (65+) per hundred children (0–14)
[c]Aged dependency ratio: aged per hundred 15–64

and subsequent labour force entrants. For example, assuming a TFR of 1.9 in the first half of this century (as well as constant mortality and zero net migration) the UK's population would be nearly 10% higher in 2050 than if its TFR was 1.7. Despite this, moderate increases in fertility would not of themselves prevent net losses from labour forces cumulating for decades.

Others aspects of generational changes, such as shifts in the relative sizes of younger and older age groups, have further implications for the ability of societies to maintain themselves. Coming decades will witness dramatic shifts in indicators of this, namely the aging index and the aged dependency ratio. Large cohorts continue to rise up the age structure 'escalator' to the older ages while below replacement fertility results in falling numbers of children. Even in the year 2000, Germany, Japan and Southern European countries already had numbers of older people equal to or exceeding the numbers of children, as shown by aging indices of 100 or more (Table 3.4). The statistics for 2000, however, only mark the beginning of the shift towards the situation where the aged outnumber children. By 2025, in Germany and Japan there could be two older people to every child, while in all the other countries in Table 3.4 the aged are also projected to outnumber children substantially – except in Australia, New Zealand and the United States. The figures for 2050 more strongly reflect consequences of long-term below replacement fertility which would deepen the trend and extend it to all countries. This would include the United States where the 2008 medium variant projections assume below-replacement fertility after 2015. Extreme outcomes in 2050, with two or three older people per child, are anticipated in Japan, Singapore, Hong Kong, Germany, Austria, parts of Eastern Europe and most of Southern Europe.

Of itself, the contracting size of the child generation is unlikely to free up sufficient extra resources – for example from the funding of primary and secondary education – to offset the extra costs of a much larger generation of the aged. An early estimate for

OECD countries indicated that per capita public expenditure on persons aged 65 and over in more developed countries was three times higher than for children under 15 (Holzmann 1988: 430). A later estimate found that a child absorbs more resources than an old person, with the result that the first 20 years of life cost more, especially in terms of private expenditure, than the total years lived after 60. Combining public and private expenditures, children appear more expensive per capita. This seems to imply that the decreased cost of supporting children is offsetting the increased cost of supporting the aged. Yet much of the decreased cost of children will be a saving in private expenditure, which the taxation system cannot easily redirect to the aged (Easterlin 1996). Some writers envisage intergenerational rivalry arising from competition for society's resources, especially in view of the mounting political weight of the elderly. Where they comprise a quarter of the total population, they will comprise a higher proportion of the electorate nationally and even a majority in certain communities of out-migration and retirement. Yet the aged are diverse in terms of age, social class, ethnicity, disposition and political allegiances; older voters have an interest not just in their own welfare but in the welfare of their children, their grandchildren and the wider community.

The so-called aged dependency ratio, despite its disadvantages (see National Research Council 2001: 42), helps to illustrate the mounting difficulty of raising sufficient tax revenue from people of working age to support the elderly (Table 3.4). Like the aging indices, aged dependency ratios are useful indicators of changes in the relative numbers in broad age groups. The ratio of older people to 'workers' is particularly relevant to issues surrounding the economic sustainability of population aging. The model of the demographic transition shows that aged dependency peaks in the post-transition stage at 25 older people per hundred of working age (see Table 1.3). As noted earlier, however, the effects of continuing demographic change, as exemplified in projections for Italy to 2050, could raise the aged dependency ratio above 60 older people per hundred 15–64. In a number of countries, the aged dependency ratios are projected to more than double in the first half of this century, with substantial increases already by 2025 (Table 3.4). Particularly high ratios, of 50 aged per 100 workers in 2050, are projected for Germany, Austria and the countries of Southern Europe. Such figures denote major changes, especially considering that many people of working age are not in the labour force. More favourable ratios are projected for the United Kingdom, the United States, Australia and New Zealand. Yet even in these countries trends in aged dependency ratios have already prompted policy interventions.

3.6 Conclusion

Age structure changes underlie the 'silent revolution' transforming societies around the world. The revolution proceeds from the flow of generations from one stage of life to another. The revolution has been long expected, but never on the scale now anticipated because of the cross-over from the demographic transition to circumstances associated with the second demographic transition. Dramatic changes are occurring in the quantum of generational flow. In the past, younger generations always outnumbered their parent's generation; now in many countries this long-established pattern is reversing. Indices of generational change and labour force change capture key consequences of shifts in age structure evolution. They provide examples of projections demonstrating futures that in a majority of cases would be better avoided. Labour force trends are a major influence on the sustainability of population aging. Even more fundamental are developments in family formation which ultimately drive much of the processes of generational change and age structure evolution. A particular challenge for the immediate future in many countries is to reshape age structure evolution through policies that are ultimately conducive to near-zero population growth, rather than to high rates of population growth or to population decline and hyper-aging.

References

Easterlin, R. A. (1996). Economic and social implications of demographic patterns. In R. H. Binstock & L. K. George (Eds.), *Handbook of aging and the social sciences* (4th ed., pp. 84–93). San Diego: Academic.

Holzmann, R. (1988). Ageing and social security costs. *European Journal of Population, 3*, 411–437.

ISTAT (Instituto Nazionale di Statistica). (2007). *Resident population projections, Italy, 2007–2050*. http://demo.istat.it/uniprev/index.html?lingua=eng. Accessed Apr 2010.

National Research Council. (2001). *Preparing for an aging world: The case for cross-national research*. Washington, DC: National Academy Press.

OECD. (1998). *Maintaining prosperity in an ageing society: Summary, June 1998*. http://www.oecd.org/dataoecd/21/10/2430300.pdf. Accessed Sept 2010.

United Nations. (2001). *World population prospects: The 2000 revision (CD, disk 1: standard set)*. New York: United Nations, Department of Economic and Social Affairs, Population Division.

United Nations. (2009). *World population prospects: The 2008 revision population database*. New York: United Nations, Department of Economic and Social Affairs, Population Division. http://esa.un.org/UNPP/index.asp?panel=2. Accessed Apr 2010.

Chapter 4
The 'New Era' in Health

Health status is one of the most important indicators of well-being, and it predicts a large proportion of societal expenditures on health and social services for the elderly.

(National Research Council 2001: 201)

4.1 New Longevity

In the last quarter of the twentieth century the population of Japan's subtropical island prefecture of Okinawa achieved prominence as having the highest life expectancy in the world, the lowest risk of major age-related chronic diseases in Japan, and the highest proportions surviving to 100. A long-standing and reliable family register system verified this (Willcox et al. 2008). More recently the prefecture's lead may have waned as younger Okinawans have adopted Western diets and forsaken traditional practices. Nevertheless the reasons underlying the longevity of the oldest generations remain of great interest and practical importance because of lessons potentially transferable to other populations. Lifestyles, rather than genetics or scientific interventions, have accounted for Okinawa's exceptional longevity. Beneficial practices include low calorie intake ("only eating until 80% full"), a diet high in vegetables, fruit and fibre, regular exercise, moderate alcohol use, and avoiding smoking. Other important influences are thought to be stress coping mechanisms associated with strong social support and deep spirituality – especially among older women – together with optimistic attitudes, adaptability and an easy-going approach to life (Cockerham and Yamori 2001: 157; Suzuki et al. 1995). The Okinawans' cardiovascular and cancer mortality rates have been up to 40% lower than the national average for Japan (Willcox et al. 2004: 789 and 793).

Among all nation states, life expectancy is also highest in Japan as a whole where, in mid-2010, the figures were 86 years for women and 79 for men (Population Reference Bureau 2010). Despite this there is a substantial gap between the average

D.T. Rowland, *Population Aging: The Transformation of Societies*, International Perspectives on Aging 3, DOI 10.1007/978-94-007-4050-1_4,

length of life and average length of life in good health. Typically, Japanese men aged 60 spend about 4 years in ill health over their remaining lifetimes, while the figure for Japanese women is 5 years. This amounts to nearly 20% of their later years – counting all periods of ill health, not just a final illness. Japan's achievement in life expectancy is evidence of the current potential for similar progress in other aging societies, and of an enduring disparity between total years lived and years in good health. This chapter reviews current theoretical perspectives on changes in survival and health in developed countries. It focuses on ideas about the characteristics of a new fourth stage of the epidemiologic transition and the nature and causes of contemporary changes in survival and health in developed countries.

4.2 The Fourth Stage

Since the 1980s, a revolution has taken place in understanding how the goals of improving health and 'adding life to years' might be accomplished. Together, these insights have led to a reformulation of the epidemiologic transition, which is concerned with trends through time in mortality rates, causes of death and the prevalence of diseases and ill health (see Chap. 1). Until the 1980s, the notion of a three-stage epidemiologic transition provided a foundation for expectations about the limits of change in survival and health. It then became apparent that developed country populations were progressing beyond Omran's assumed limits. Clear evidence was the continuing rise in life expectancy at birth. Whereas 75 years was once thought to be a limit in the epidemiologic transition, female life expectancies of between 80 and 85 years are now widespread. They occur in most countries of Western and Northern Europe, together with the United States, Canada, Australia and New Zealand. Equally striking are the associated gains in life expectancy at age 65 where, for females, a gain of 8 years accompanies the shift from 70 to 85 years in life expectancy at birth.

Recognition of continuing changes in 'post-transition' societies prompted the suggestion that 'a new era in epidemiologic history' had begun, dating from the mid-1960s in the United States: Olshansky and Ault (1986) called it "the age of delayed degenerative diseases". They distinguished this 'new era' or fourth stage of the epidemiologic transition from the third stage by the rapidly declining death rates at older ages and a shift to even older ages in the distribution of mortality from degenerative diseases. Latitude for progress in preventing or delaying diseases long existed, for example, where early onset was related to lifestyles or to environmental influences such as air pollution.

Emergence of the current stage of the epidemiologic transition in developed countries comprises part of the new demography that is reshaping prospects for aging societies. Ideas about the nature of the fourth stage, however, have changed and broadened over time. A new assessment is now needed of the characteristics that distinguish the current stage of the epidemiologic transition in developed countries. Ensuing sections discuss the characteristics that are most important in relation to the study of aging, including the extension of life into the eighth and ninth decades, which is a key feature of the fourth stage.

4.3 Longevity

Long life is now an achieved goal for many countries and gains are still occurring. Recent reductions in later life mortality are unprecedented and unexpected (Christensen et al. 2009: 1196). The phenomenon known as 'the rectangularization of the survival curve' summarizes the revolution in survival in middle and later ages that that has been characteristic of the epidemiologic transition (Fig. 4.1). It provides broad evidence of a long-term trend towards a narrower range in the ages at which most deaths occur, particularly because of the decline in death rates of infants, children and young adults.

Figure 4.1 shows the changing shape of a population's survival curve, which represents the proportions surviving to successive ages from an original 100,000 live births. The curve becomes more rectangular as infant and child mortality decline and higher proportions survive to maturity and later ages. In the fourth stage of the epidemiologic transition, rectangularization becomes especially pronounced because the longest life expectancies at birth entail close to 100% survival of females to their 50th birthday (i.e. 'exact age' 50). At the same time there occur great increases in the proportions reaching older ages. Thus female life expectancies at birth of 70, 80 and 90 are associated respectively with survival to age 75 of 52%, 76% and 94%. Some countries already have around a third of their female population reaching their mid-eighties. A rise in this figure to 80%, assuming life expectancy at birth reaches 90, would mean that a protracted period of dependency had become a typical part of the life course.

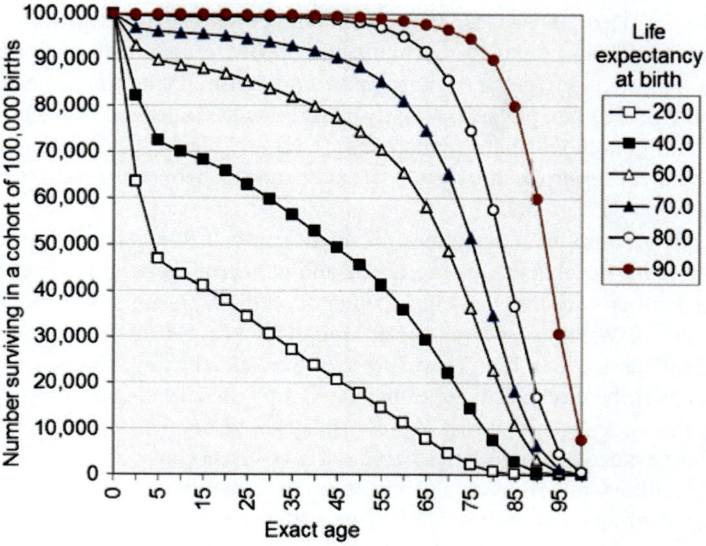

Fig. 4.1 Rectangularization of the survival curve for females (Source: Coale and Demeny (1983) and author's calculations for a life expectancy of 90)

Prolonged survival, and longer survival through ages at greatest risk of frailty, are characteristic of the current stage of the epidemiologic transition. Progress from a life expectancy of 80 to 85 and from 85 to 90 entail major increases in the number of years individuals live in later life – because there is little remaining scope for further improvements in survival among the young. Late in the epidemiologic transition, increases in life expectancy at birth are due especially to longer life at older ages, to an extent greater than any earlier changes at these ages.

Oeppen and Vaupel (2002: 1029) argue that life expectancy is not yet approaching its maximum, particularly because of improvements in adult survival. The authors assert that all previous estimates of the limits to human life expectancy have been shown to be false within 5 years of publication. They note that, in the best circumstances, life expectancy has increased by 2.5 years per decade for a century and a half; a continuation of this, they envisage, would raise the record life expectancy from the current 86 years for females in Japan to 100 years by about 2060. In other words "centenarians may become commonplace within the lifetimes of people alive today" (ibid.: 1031). No limit is in sight in the longest lived populations. This is raising the possibility that a majority of babies born in the first decade of the twenty-first century will celebrate their 100th birthdays (Christensen et al. 2009: 1196). Such a development would greatly augment the role of mortality decline as a force in population aging.

The current revolution in survival is due to a range of factors including the important role of moderate interventions, such as raising awareness of distinctions between healthy and unhealthy lifestyle practices, screening for early detection of common diseases, and prescription of medications to control diseases and their and precipitating factors. The decline in death rates from diseases of the circulatory system, such as coronary heart disease and stroke, has been particularly important (Andrews 2001: 6). Disease prevention, delayed onset, death prevention from particular causes and delayed mortality all contribute to longer survival overall. The *National Strategy for an Ageing Australia* identified smoking, poor nutrition, alcohol and inadequate physical activity as significant contributors to the burden of disease, and concluded that "perhaps as much as half of the functional decline associated with aging is the result of disuse and can be reversed by exercise" (ibid.: 40 and 43).

A negative consequence of more prevalent survival to advanced ages has been expansion in the numbers with severely disabling neurodegenerative diseases, especially Alzheimer's disease, vascular dementia and Parkinson's disease (Broe and Creasey 1995). Without research breakthroughs in preventing or treating neurodegenerative diseases, their high prevalence will remain a leading characteristic of the fourth stage of the epidemiologic transition, simply because higher proportions are reaching the ages at greatest risk. Appropriate care of dementia patients, especially those with advanced symptoms, is particularly costly because it calls for dementia-specific facilities rather than placement in mainstream nursing homes (Senate Community Affairs Committee 2005: page 5.25).

4.4 Hubris

Another suggested feature of the fourth stage of the epidemiologic transition consists of engaging in hubristic behaviour, ignoring risks and endangering one's own health or life. Rogers and Hackenberg saw this as a key characteristic, dating from the 1970s in the United States. They argued that: "death is becoming increasingly influenced by individual behaviours and new life-styles" (Rogers and Hackenberg 1987: 239). Risk-taking has many forms including smoking, alcohol abuse, avoiding exercise, eating too much and consuming foods high in saturated fat, salt and sugar. For example, four out of five older Australians have chronic diseases influenced by diet (Volker and Caterson 2001, cited by Temple 2006: 28). Other risks are driving at high speeds, using addictive drugs, sharing injecting needles and engaging in unsafe sex. All are behaviours entailing health risks of which many are aware, but may choose to ignore. Rogers and Hackenberg (1987: 240) also included accidents, suicides, homicides, avoidance of vaccinations for preventable diseases, and the rise in lifestyle-related cases of HIV/AIDs as instances of hubristic behaviour. Although the consequences of some forms of hubristic behaviour occur early or immediately, the inability to modify unhealthy behaviours in the long term can be fatal in later years.

Smoking is one of the most devastating forms of hubristic behaviour in developed countries. A 50 year study of British doctors, which commenced in 1951, found that doctors who smoked throughout their lives died, on average, 10 years younger than lifelong non-smokers; their habit killed more than half the lifelong smokers. Stopping smoking at age 50 halved the mortality, while stopping at age 30 avoided most of the risk (Doll et al. 2004: 1519). The authors concluded that it is reasonable to assume that, in the future, half the smokers world-wide will die of smoking related diseases, although cessation, even in mid-life, offers hope for a substantial catch-up in survival (ibid.: 1528). Stampfer (2004: 1507), observed that smoking commences at much younger ages today, implying that the current generation of persistent smokers will face even higher excess mortality, most conspicuously where the prevalence of smoking has increased – in developing countries, among women and lower socio-economic groups.

A focus on hubristic behaviour emphasizes individual agency – people's responsibility for their own health. It implies an approach to health policies in which promotion of lifestyle changes is a key aspect. Another viewpoint argues that people's position in society needs to be considered as well: "Blaming the victim fails to address underlying questions of why disadvantaged people are drawn into these behaviours and the nature of the social and individual influences that maintain them" (Jarvis and Wardle 1999: 241). Hubristic behaviour can scarcely be ascribed to people who have limited choices.

Accordingly, individuals are not solely responsible for their own lifestyles and health. What some may perceive as outcomes of reckless or thoughtless behaviour, others regard as outcomes of the social context in which people live. An influence on health that arises from the nature of the society is whether the health system provides timely, affordable and adequate care. Deficiencies in health care provision can lead, not to general disadvantages throughout the society, but to social inequalities such that the most deprived experience the most adverse consequences. For example, under-funding and under-supply of health care can have greater impacts on the poor because they are least able to afford health insurance, dentistry, medications, consultations with specialists and early treatment in a hospital. Social class further influences people's health if they cannot afford good housing or good nutrition, or if there is peer-group or advertising pressure for smoking, excessive alcohol consumption and other risky behaviours. Some argue, therefore, that health inequalities need to be addressed particularly through policies that address education, income and other aspects of socio-economic inequality (Kawachi et al. 1999).

4.5 The Status Syndrome

The influence of social class is more pervasive than evident from broad distinctions between rich and poor. Since the 1980s an important realization has been that people's location at seemingly advantaged levels in the social hierarchy can be a health hazard. This is because of the stresses associated, not with economic deprivation itself, but with individuals' lack of control over their own circumstances and lack of opportunities for full social participation: "control over one's life and opportunities for meaningful social engagement are necessary for health" (Marmot 2004b: 241). This creates 'the social gradient of health' or 'the status syndrome' (Marmot 2004b; Marmot and Wilkinson 2006: 2). The social gradient concept, focusing on social inequality and health, originated from the 'Whitehall studies' of British civil servants. The research revealed that even though the subjects were neither very wealthy nor very poor, the lower their position in the civil service hierarchy, the greater the death rate (Marmot 1999: 10–12). Lifestyle factors explained less than a third of the social gradient, which prompted the conclusion that other social and psychosocial factors – linked to social class – were major influences.

Higher stress was found to be associated with lower social status for reasons arising in the social environment in childhood, as well as in the work environment and the community. Long term adverse effects on health, with high risk of premature illness and death, were thought to arise from childhood deprivation, such as being raised in stressful circumstances due to poverty and poor diet, possible consequences of which are impaired physical and mental development. In employment, ill health was linked with lack of control over work, lack of variety and small chance of reward, leading to chronic anxiety and low self-esteem – which undermine mental and physical health. In contrast, 'executive stress' in the uppermost social rank, where high effort was linked to high reward, was thought to be "satisfying and generally health promoting". Other potential causes of the social gradient of health were insecurity of employment, unemployment, shame, inferiority, subordination, social isolation, difficulty paying bills, distressing life events and exposure to hostility within communities (Brunner and Marmot 1999: 26 and 32; Wadsworth 1999: 51; Wilkinson 1999: 260).

Studies of animal groups with hierarchical social arrangements have provided some corroborative evidence as well as indicating a biological mechanism. Among baboons, subordinate social status can induce chronic anxiety leading to increased arteriosclerosis, high cholesterol, abdominal obesity, depression and poorer immune function. A biological mechanism here is the 'stress hormone' cortisol which, if maintained at high levels, is harmful to health. Baboons well illustrated the health effects of social status independently of lifestyle risk factors (see Wilkinson 1999: 260; Brunner and Marmot 2006: 26).

Research on the social gradient has implications for older people because the risk of relatively early death in later life is associated with the number of years lived in disadvantaged social positions. People take into later life the accumulated damage to their health incurred earlier. The social gradient of health envisages that adverse health consequences flow from relative deprivation in the social hierarchy, leading to stress-induced damage implicated in cardiovascular disease, cancer, infection, cognitive decline and accelerated aging. In developed countries, relative rather than absolute deprivation is paramount, with health related less to people's absolute material living standards than to their position in society (Blane 2006; Brunner and Marmot 2006).

Box 4.1 Social Inequality and Smoking

Although lifestyle risk factors are easier to address than the social gradient, a focus on risk factors alone can preserve or widen health inequalities, because lifestyle advice and controls may be least helpful to the most deprived groups. The risk factor of smoking illustrates this point. Jarvis and Wardle's (1999, 2006) research in Britain in 1996 found that there was a social gradient in smoking, with 80% of the poorest fifth of the adult population smoking compared with less than 20% of the most affluent fifth. Also, although the prevalence of cigarette smoking had declined steeply over the previous 20 years, smoking rates fell particularly among the affluent, with little change for the most deprived. The gulf in smoking prevalence between social groups has continued to widen.

There was a similar social gradient in death rates from smoking-related diseases, including heart disease, lung cancer, emphysema and chronic airways obstruction. The British evidence pointed to similar desires to quit smoking across all social groups, but with nicotine dependence increasing with deprivation arising from smoking more cigarettes more intensively. The hypothesis that deprived people smoke to relieve stress was not supported. Rather, deprived people smoked more because of a need to relieve their higher levels of nicotine dependence and stronger withdrawal symptoms when not smoking. The authors argued that successful quitting leads to lower rather than higher levels of perceived stress. For deprived people with already more

(continued)

Box 4.1 (continued)

stressful lives, contending with an extended period of nicotine withdrawal appears to make quitting smoking all the more difficult. Increasing the price of cigarettes, health education campaigns, and banning smoking in public places do not address the problem of nicotine dependence. While such interventions have reduced the overall prevalence of smoking, they have increased health inequalities as well as economic inequalities because those most dependent on cigarettes and least able to afford them have continued to smoke. Some suggested strategies to address the social gradient are nicotine replacement therapy, reimbursement of the cost of smoking cessation treatments, and product development to make smoking less harmful (Jarvis and Wardle 1999, 2006).

Accordingly, while some have argued that the most effective way to reduce excess mortality from diseases such as coronary heart disease is through population-wide control of risk factors – including high cholesterol, high blood pressure, smoking and inactivity – others point out that this approach ignores the additional influence of social inequality (Box 4.1). In relation to coronary heart disease in Britain, population-wide strategies have brought the greatest improvement among people of higher socioeconomic position, thereby widening social inequality (Marmot 2004a: 297–298). This has led to recognition that population-wide control of major risk factors needs to give particular attention to the most vulnerable groups (Emberson et al. 2004: 1153). In risk factor modification there is a major role for the primary health care sector – general practitioners, community nurses, pharmacists, physiotherapists and other allied health professionals– who have opportunities to make early interventions and promote preventative strategies and behaviours (Andrews 2001: 50).

4.6 The Paradox of Health

A further potential impediment to health improvements in the fourth stage of the epidemiologic transition is fear of illness. This first came to prominence in a paper on 'the paradox of health' in the *New England Journal of Medicine* (Barsky 1988). Barsky observed that respondents in American health surveys were reporting a higher incidence of illnesses, a trend associated with (a) an increased prevalence of chronic and degenerative disorders; (b) society's heightened consciousness of health leading to greater self-scrutiny and amplified awareness of bodily symptoms and feelings of illness; (c) a climate of apprehension, insecurity and alarm about disease created by commercialization of health and media attention to health issues; (d) the progressive medicalization of everyday life, bringing unrealistic expectations of cure that make untreatable infirmities and unavoidable ailments seem worse. Barsky (1988: 415) considered there was occurring:

> … a progressive decline in our threshold and tolerance for mild disorders and isolated symptoms, along with a greater inclination to view uncomfortable symptoms as pathologic – as signs of disease. Coupled with this has been a readiness to adopt the "sick" role, to seek medical care for isolated symptoms, and to acknowledge to others that we feel ill. The standard we use for judging our health appears to have been raised, so that we are more aware of – and more disturbed by – symptoms and impairments that previously we deemed less important.

Barsky's paradox raises the possibility of adverse consequences from health promotion and publicity about diseases. The paradox of health implies a widening gap between objective health status and subjective well-being (Barsky 1988: 414). Rowe and Kahn (1998: 25–26), for example, observed that: "Older people have become so sensitized to the threat of Alzheimer's disease that every forgotten name or misplaced key ring strikes fear." People in ill health, however, are not equally affected by medical concerns. Some may be "body transcendent" while others are "body preoccupied", leading to differences in activity levels and well-being among people with the same objective health status (Shanas et al. 1968: 67). The broad implication is that health promotion programmes should seek not only to encourage risk avoidance and early recognition of diseases, but also to foster well-being and acceptance of the inevitability of biological aging.

4.7 Health Risk Factors

Bombardment with information about the symptoms of dread diseases needs to be counterbalanced with positive information about health-protective behaviours that can foster long-term benefits. Although early recognition of symptoms saves lives, health practices that enable avoidance of disease symptoms can save many more. This is essential because a relatively small number of preventable or controllable health risks are responsible for a substantial proportion of deaths and lost years of healthy life. Three types of factors comprise the main health risks in high income countries (e.g. in North America, Northern, Western and Southern Europe, and Australasia), namely addictive substances (tobacco and alcohol) and factors related to diet and physical inactivity.

In high income countries the six leading causes of lost years of healthy life are, in descending order: use of tobacco and alcohol, overweight and obesity, high blood pressure, high blood glucose and physical inactivity (Table 4.1). These also account for a substantial proportion of deaths. All are risk factors for chronic diseases, such as heart disease and cancers. Tobacco use takes the greatest toll both in terms of its involvement in lost years of healthy life (11%) and total deaths (18%). The figures in Table 4.1 cannot be added, however, because many deaths are the result of more than one risk factor. The World Health Organisation (WHO) estimated that, in high

Table 4.1 Ten leading risk factors in causes of lost years of healthy life (DALYs) and death, high and middle income countries, 2004

High income countries[a]			Middle income countries[b]		
Risk factor	Percentage of total DALYs[c]	Percentage of total deaths	Risk factor	Percentage of total DALYs[c]	Percentage of total deaths
Tobacco use	10.7	17.9	Alcohol use	7.6	6.4
Alcohol use	6.7	1.6	High blood pressure	5.4	17.2
Overweight and obesity	6.5	8.4	Tobacco use	5.4	10.8
High blood pressure	6.1	16.8	Overweight and obesity	3.6	6.7
High blood glucose	4.9	7.0	High blood glucose	3.4	6.3
Physical inactivity	4.1	7.7	Unsafe sex	3.0	
High cholesterol	3.4	5.8	Physical inactivity	2.7	6.6
Illicit drugs	2.1		High cholesterol	2.5	5.2
Occupational risks	1.5	1.1	Occupational risks	2.3	
Low fruit and vegetable intake	1.3	2.5	Unsafe water, sanitation, hygiene	2.0	
Urban outdoor air pollution		2.5	Low fruit and vegetable intake		3.9
			Indoor smoke from solid fuels		2.8
			Urban outdoor air pollution		2.8

Source: WHO (2009a: 11–12)

[a] High income: gross national income per capita US$10,066 or more (World Bank definition)

[b] Middle income: gross national income per capita US$ > 825 < 10,066 (World Bank definition)

[c] DALYs: disability-adjusted life years, or lost years of healthy life (for details see Chap. 5)

income countries in 2004, 21% of lost years of healthy life and 28% of deaths were attributable to the ten risks shown in each column of the table (WHO 2009a: 5 and 30).

The potentially dire consequences of each risk factor highlight the universal benefits of health protective behaviours. Physical inactivity increases the risk of cardiovascular disease, breast and colon cancers and type 2 diabetes. It also excludes possible benefits of exercise such as improved musculoskeletal health, control of body weight and reduced symptoms of depression. Similarly, increasing levels of overweight and obesity are associated with greater risk of stroke, cardiovascular disease, type 2 diabetes, and cancers of the breast, colon, prostate and other organs. High blood pressure brings heightened risks of stroke, heart disease, kidney failure and other diseases. It is commonly caused by the effects of too much salt, as well as alcohol, obesity and lack of exercise. High cholesterol increases the risks of heart disease, stroke and other vascular diseases. High blood glucose is related to diet and physical inactivity which increase resistance to insulin, but genetics also have an important role. Globally raised blood glucose causes all deaths from diabetes, as well as 22% of deaths from ischemic heart disease and 16% from strokes. Consuming fruit and vegetables may prevent some cancers but the main benefit is from reducing cardiovascular disease. Whereas regular exercise is one the most beneficial interventions individuals can make for their own health, smoking is one of the worst. It greatly increases the risk of dying from a range of diseases, including lung and other cancers, heart disease, stroke and chronic respiratory disease. Alcohol contributes to more than 60 types of disease, such as oesophageal cancer and liver cancer, as well as to motor vehicle accidents, other injuries and homicide. In relation to cardiovascular disease in older people, its net effect may be protective provided consumption is moderate. The WHO estimated that ischemic stroke deaths would be 11% higher in high-income countries if no one drank alcohol. Despite this the overall impact of alcohol is harmful (ibid.: 16–21).

The above summary by no means captures all of the ramifications of the leading risk factors and it omits other risk factors with health implications in later life. The WHO estimated that 33% of deaths world-wide in 2004 were attributable to ten leading risks, while 44% were attributable to 24 leading risks. If the 24 risks were eliminated, global life expectancy would have been 9 years longer, compared with 7 years without the ten leading risks (ibid.: 29–30). The relative importance of health risks changes through time and currently varies according to national income levels. The WHO uses the income classification of the World Bank. The 'middle income' group is very broad and includes countries with a range of different levels of development. For this reason it is sometimes split into 'upper middle' and 'lower middle' groups (see Table 5.1). In middle income countries, which tend to have younger populations, the percentages experiencing the consequences of the leading risk factors are generally lower but the numbers of people affected are much larger (ibid.: 11). Prominent also in middle income countries are some risk factors associated with lower levels of socio-economic development, such as unsafe water and indoor smoke (Table 4.1). In aging societies the health risks associated with environmental hazards, poor infrastructure and poverty are usually greatly reduced, but reduction of further risks is an ongoing priority.

In all aging societies hubris and social inequality persist as key obstacles to the dissemination of knowledge of health risk factors and empowerment to act upon such knowledge. In addition, the most disadvantaged socio-economic groups are more likely to have low levels of health literacy, which the WHO (1998: 10) defined as: "the cognitive and social skills which determine the motivation and ability of individuals to gain access to, understand and use information in ways which promote and maintain good health." Health literacy has applications in addressing health risks through its recognition of a range of relevant strategies. These include enhancing access to and comprehension of information about protective behaviours, improving decision making in the event of illness, enabling effective use of available health services, facilitating understanding of medical advice, and obtaining culturally appropriate care (see Morrow et al. 2006; Sadowski 2010). Implementation involves responsiveness of health professionals and the health system to particular needs and circumstances as well as motivation and willingness to learn on the part of individuals. Australian survey data show that the percentages aged 15–74 with at least an adequate level of health literacy ranged from 26 in the lowest socio-economic status areas to 55 in the highest (AIHW 2010: 81). More striking than the socio-economic differences, however, was the low percentage of Australia's total population with adequate health literacy: only 41%. This was despite Australia's high living standards and a GDP per capita ranking eighth in the OECD (OECD 2010a). Continuing low levels of health literacy is a likely further characteristic of the fourth stage of the epidemiologic transition, one which may partly account for hubristic behaviour.

The prevalence of health risks and premature mortality are concerns in most countries, although some with the lowest birth rates – Italy, Japan and Singapore – are relatively well-positioned in relation to health risks (Table 4.2). Among Western countries, only the United States has premature mortality in excess of 10% of a birth cohort dying between the ages of 15 and 60. High death rates of American males are responsible for this anomaly: 14% of males die between these ages, compared with 8% of females. Meanwhile the Eastern European countries have losses at the same ages ranging from 10% to 30%, placing some well above the level for China and even India.

The most important factors in Eastern Europe's level of premature mortality are death rates from cardiovascular diseases for which the highest levels occur in the Russian Federation and Ukraine. Cancer mortality rates are also relatively high in Eastern Europe (Table 4.2). In the developing countries the prevalence of tobacco use, especially by males, is indicative of a future major burden of smoking-related disease. Most of the lowest rates of tobacco use by males occur in North America and Australasia but even there the proportions smoking remain a significant health concern for governments. Similarly, although smoking is not as prevalent among females in all the countries in Table 4.2 the proportion of the population involved is substantial. Obesity is a relatively new 'epidemic' in which the United States leads with over 30% of men and women having BMI's of 30 or more. Australia, New Zealand, the United Kingdom and Greece are the only other Western countries in the table with more than 20% obese. Finally, while the WHO's data show that alcohol consumption of 5–10 L of pure alcohol per capita is typical in many countries the figures do not reveal peak figures, for instance where men, or a proportion of them, consume much of the alcohol.

Table 4.2 Mortality and health risks in selected countries, 2000–2007

	Proportion dying at ages 15–60 per 1,000	Age-standardized mortality rate for cardiovascular diseases per 100,000	Age-standardized mortality rate for cancer per 100,000	Percentage using tobacco at ages 15 and over	Percentage using tobacco at ages 15 and over	Percentage obese at ages 15 and over	Percentage obese at ages 15 and over	Per capita alcohol consumption (litres of pure alcohol) ages 15 and over
Sex:	Both	Both	Both	Males	Females	Males	Females	Both
Year:	2006	2002	2002	2005	2005	2001–2007	2001–2007	2003
North America and Australasia								
United States	109	188	134	26.3	21.5	31.1	33.2	8.6
Canada	72	141	138	24.3	18.9	15.9	13.9	7.8
Australia	65	140	127	27.7	21.8	22.3	20.4	9.0
New Zealand	75	175	139	29.7	27.5	21.9	23.2	9.7
Northern and Western Europe								
United Kingdom	80	182	143	36.7	34.7	22.3	23.0	11.8
France	91	118	142	36.6	26.7	10.5	10.4	11.4
Switzerland	63	142	116	30.7	22.2	7.9	7.5	10.8
Belgium	86	162	148	30.1	24.1	11.9	13.4	10.6
Netherlands	70	171	155	38.3	30.3	10.2	12.2	9.7
Denmark	88	182	167	36.1	30.6	9.8	9.1	11.7
Norway	70	181	137	33.6	30.4	6.4	5.9	5.5
Sweden	64	176	116	19.6	24.5	10.4	9.5	6.0
Finland	96	201	115	31.8	24.4	14.9	13.5	9.3
Germany	81	211	141	37.4	25.8	13.6	12.3	12.0
Austria	79	204	127	46.4	40.1	12.0	12.7	11.1
Southern Europe								
Italy	64	174	134	32.8	19.2	7.4	8.9	8.0
Greece	76	258	132	63.6	39.8	26.0	18.2	9.0
Slovenia	104	228	160	31.8	21.1	16.5	13.8	6.7

(continued)

Table 4.2 (continued)

	Proportion dying at ages 15–60 per 1,000	Age-standardized mortality rate for cardiovascular diseases per 100,000	Age-standardized mortality rate for cancer per 100,000	Percentage using tobacco at ages 15 and over	Percentage using tobacco at ages 15 and over	Percentage obese at ages 15 and over	Percentage obese at ages 15 and over	Per capita alcohol consumption (litres of pure alcohol) ages 15 and over
Sex:	Both	Both	Both	Males	Females	Males	Females	Both
Year:	2006	2002	2002	2005	2005	2001–2007	2001–2007	2003
Spain	75	137	131	36.4	30.9	13.0	13.5	11.7
Portugal	93	208	140	40.6	31.0	14.6	16.1	11.5
Eastern Europe								
Poland	145	324	180	43.9	27.2	15.7	19.9	8.1
Czech Republic	108	315	177	36.6	25.4	13.7	16.3	13.0
Croatia	113	356	167	38.9	29.1	21.6	22.7	12.3
Bulgaria	157	554	125	47.5	27.8	–	–	5.9
Russian Fed.	300	688	152	70.1	26.5	–	–	10.3
Ukraine	264	637	139	–	–	–	11.3	6.1
Hungary	177	364	201	45.7	33.9	17.1	18.2	13.6
Latvia	223	482	156	54.4	24.1	11.9	19.5	9.6
Asia and Latin America								
Japan	67	106	119	44.3	14.3	2.9	3.3	7.6
Singapore	67	171	128	–	–	6.4	7.3	2.2
China	116	291	148	59.5	3.7	2.4	3.4	5.2
Indonesia	212	361	132	65.9	4.5	1.1	3.6	0.1
India	241	428	109	33.1	3.8	1.3	2.8	0.3
Bangladesh	254	428	111	47.0	3.8	–	–	0.0
Pakistan	206	425	107	35.4	6.6	–	–	0.0
Mexico	122	163	88	36.9	12.4	18.6	28.1	4.6
Brazil	176	341	142			8.9	13.1	5.8

Sources: WHO (2009b) and OECD (2010a, b)

4.8 Compression of Mortality and Morbidity

The most positive evidence of change in the fourth stage is the continuing rise in life expectancy. Counterbalancing this are a range of influences such as: Barsky's paradox, the status syndrome, hubris, the rising prevalence of neurodegenerative diseases, the high prevalence of health risks and low levels of health literacy. Mixed outcomes for societies are therefore apparent. Nonetheless, there is considerable interest in seeking evidence of whether an overall improvement in the health of aging populations is accompanying the broad rectangularization of the survival curve, and whether rectangularization is actually resulting in a greater concentration of deaths in the oldest ages. Potential indicators of such developments are measures of 'the compression of mortality' and 'the compression of morbidity'. The majority of people do not die at or near the average life expectancy, which is merely a summary statistic for a range of outcomes above and below it. In long-lived populations, more than half live beyond the average life expectancy at birth. Also, although most deaths now occur after age 50, an important feature of present and projected future mortality is the considerable spread in the ages at which deaths occur, as illustrated in Fig. 4.2. A key question is whether this spread is narrowing over time, in other words is a 'compression of mortality' occurring?

The concept of compression of mortality is best known from the work of James Fries (1980) who considered that the upper limit of human life expectancy is about 85 years and, since many deaths occur in old age, there would be a tendency for mortality to become 'compressed' around age 85 (Wilmoth and Horiuchi 1999: 476).

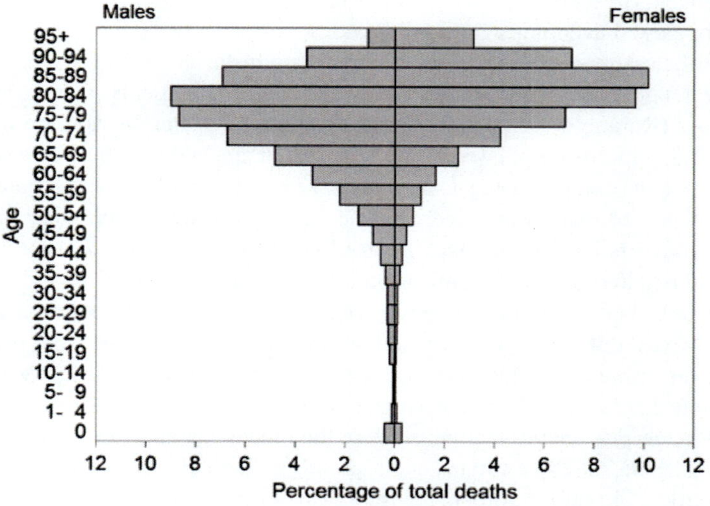

Fig. 4.2 Distribution of deaths in a long-lived population (*Note*: The first two age groups are <1 and 1–4 years) (Source: 'West' model life table with life expectancies of 80.0 for females and 76.6 for males)

The ultimate outcome would be near universal survival until people reach their eighties, followed by a concentration of deaths around the supposed biological limit. Fries (1980) also introduced the term 'natural death' referring to death at the biological limit to life, that is, in the absence of exogenous or socially avoidable influences.

A substantial decline in the variability of ages at death occurred between the 1870s and 1950s, as death rates at young ages fell (Wilmoth and Horiuchi 1999: 475). Thus there is evidence of compression in the general shift towards older ages at death. However, there is less agreement about compression of mortality within age groups 65 and over. An opposing view, predating Fries hypothesis, is that of Gruenberg (1977) who concluded that life-preserving technology is merely keeping sick people alive longer – the opposite of Fries' belief in a universal extension of life in good health. Gruenberg's hypothesis about 'the failures of success' however, gives too little recognition to the contribution of medical advances to preventing diseases and disabilities and successfully treating formerly untreatable conditions (Howse 2006: 5). An early assessment of Fries' hypothesis found that there was greater variability in the ages at death of older people, rather than decreasing variability due to a compression of mortality around some age representing a natural life span (Myers and Manton 1984). Supporting this finding was an accompanying analysis of the age distribution of deaths from five major causes, thus disproving the notion that mortality compression might be pronounced for particular causes of death. Nor could the authors find, from authoritative actuarial sources, evidence that the United States population was currently near a biological limit to life span that was constraining increases in life expectancy (ibid.: 348). Later research similarly found that neither mortality nor morbidity was becoming compressed at older ages (Wilmoth and Horiuchi 1999: 475). For example, groups with higher socio-economic status appear to have longer life in ill health.

Nevertheless more recent research employing historical data, for nine countries with the longest life expectancies, produced qualified evidence supporting Fries' hypothesis (Robine et al. 2008a). The study found that the distribution of ages at death became more compressed around the modal age at death, but there was some evidence of the trend ceasing as countries reached higher life expectancies. The authors observed that, in the nine countries, "a significant compression of old-age mortality occurred during the last 50 years." The general pattern was that "the higher the modal length of life, the more concentrated the distribution of the individual life spans" (ibid.: 11). Contrary to Fries' theory, however, they noted a continuing shift to longer lives, rather than compression around 85 years, as well as the persistence of a greater dispersion in the ages at death than Fries envisaged (ibid.: 5–10). They also found that, among Japanese women, compression of mortality ceased in the late 1980s, despite continuing increases in the modal age at death.

Fries (1980: 28–29) estimate of 85 as the biological limit of life raises other complexities, including the problem of differentiating between endogenous (biological) and exogenous (external) causes of death and calculating their respective contributions to overall mortality. Exogenous causes are supposedly preventable causes including social and environmental factors. Yet distinctions between the two types of causes are often arbitrary: the factors responsible for 'natural death' are impossibly intertwined with exogenous causes (Olshansky and Ault 1986: 381).

The 'compression of morbidity' hypothesis, associated with the notion of compression of mortality, envisages that later life morbidity will ultimately become compressed into the period immediately before death. Its realization would require not only the elimination of all exogenous factors in mortality but also elimination of the chronic disabling diseases and conditions that are most prevalent in later life. Genetic differences in susceptibility to disease militate against morbidity compression, and Gruenburg's 'failures of success' suggests a lengthening of the average duration of ill health among the aged. Although extension of life into very old ages, together with earlier and more comprehensive detection and treatment of diseases, supports Gruenburg's hypothesis, interventions that reduce the severity of diseases and prolong survival imply success rather than failure. Fries' defence of his hypothesis argues that compression of morbidity is possible if primary prevention substantially delays the onset of disabling diseases (Howse 2006: 7). This would require sufficient delay to counterbalance the effects of still longer life expectancies and more person-years lived at ages 85 and over. Currently, there is no consensus about a shift towards a compression of morbidity. There are thought to be different trends for different diseases and conditions, as well as variations between countries (Robine et al. 2008b; Cox and Hope 2006: 232). Also "the kind of evidence that is needed to support solid conclusions about the expansion or contraction of morbidity is simply not available for most countries in the developed world" (Howse 2006: 19).

4.9 Conclusion

For older age groups, as well as society generally, key characteristics of the 'new era' in health, or the current stage of the epidemiologic transition, are features associated with demographic changes, and features arising from the social context of people's lives, namely: (i) continuing improvements in life expectancy, now due especially to better survival in middle and later life, which inevitably raises the proportions reaching the ages where chronic illnesses and disabling diseases have their highest prevalence; (ii) a decline in mortality from certain degenerative diseases, because of prevention, delayed onset or earlier diagnosis and treatment; (iii) a great increase in the prevalence of degenerative diseases of the brain as higher proportions survive to advanced ages; (iv) a more conspicuous role for hubristic behaviour accompanied by the persistence of health risks and inadequate health literacy, despite better knowledge and higher levels of education; (v) the pervasive influence of social inequality on health; (vi) negative, as well as positive, effects of the medicalization of everyday life.

References

AIHW (Australian Institute of Health and Welfare). (2010). *Australia's health 2010: The twelfth biennial health report of the Australian institute of health and welfare* (Cat. No. AUS 122). Canberra: Australian Institute of Health and Welfare.

Andrews, K. (2001). *National strategy for an ageing Australia: an older Australia, challenges and opportunities for all*. The Hon Kevin Andrews MP, Minister for Ageing. Canberra: Commonwealth of Australia.

Barsky, A. J. (1988, February). The paradox of health. *The New England Journal of Medicine, 18*, 414–418.

Blane, D. (2006). The life course, the social gradient, and health. In M. Marmot & R. G. Wilkinson (Eds.), *Social determinants of health* (pp. 54–77). Oxford: Oxford University Press.

Broe, G. A., & Creasey, H. (1995). Brain ageing and neurodegenerative diseases: A major public health issue for the twenty-first century. *Perspectives in Human Biology, 1*, 53–58.

Brunner, E., & Marmot, M. (2006). Social organization, stress, and health. In M. Marmot & R. G. Wilkinson (Eds.), *Social determinants of health* (2nd ed., pp. 6–30). Oxford: Oxford University Press.

Christensen, K., Doblhammer, G., Rau, R., & Vaupel, J. W. (2009). Ageing populations: The challenges ahead. *Lancet, 374*, 1196–1208.

Coale, A. J., & Demeny, P. (1983). *Regional model life tables and stable populations*. New York: Academic Press.

Cockerham, W. C., & Yamori, Y. (2001). Okinawa: An exception to the social gradient of life expectancy in Japan. *Asia Pacific Journal of Clinical Nutrition, 10*(2), 154–158.

Cox, M., & Hope, S. (2006). Health service needs and labour force projections. In J. Boston & J. A. Davey (Eds.), *Implications of population ageing: Opportunities and risks* (pp. 221–251). Wellington: Institute of Policy Studies, Victoria University of Wellington.

Doll, R., et al. (2004). Mortality in relation to smoking: 50 years' observations on male British doctors. *British Medical Journal, 328*(26 June), 1519–1528.

Emberson, J. R., Whincup, P. H., Morris, R. W., & Walker, M. (2004). Reducing social inequalities and the prevention of coronary heart disease. *International Journal of Epidemiology, 33*(5), 1152–1153.

Fries, J. (1980). Aging, natural death, and the compression of morbidity. *The New England Journal of Medicine, 303*(3), 130–135.

Gruenberg, E. M. (1977). The failures of success. *The Milbank Memorial Fund Quarterly/Health and Society, 55*, 3–24.

Howse, K. (2006). *Increasing life expectancy and the compression of morbidity: A critical review of the debate* (Working Paper No 206). Oxford: Oxford Institute of Ageing.

Jarvis, M., & Wardle, J. (1999). Social patterning of individual health behaviours: The case of cigarette smoking. In M. Marmot & R. G. Wilkinson (Eds.), *Social determinants of health* (pp. 240–255). Oxford: Oxford University Press.

Jarvis, M. J., & Wardle, J. (2006). Social patterning of individual health behaviours: The case of cigarette smoking. In M. Marmot & R. G. Wilkinson (Eds.), *Social determinants of health* (2nd ed., pp. 224–237). Oxford: Oxford University Press.

Kawachi, I., Kennedy, B. P., & Wilkinson, R. G. (Eds.). (1999). *The society and population health reader: Income inequality and health*. New York: New Press.

Marmot, M. (1999). Introduction. In M. Marmot & R. G. Wilkinson (Eds.), *Social determinants of health* (pp. 1–16). Oxford: Oxford University Press.

Marmot, M. (2004a). Commentary: Risk factors or social causes? *International Journal of Epidemiology, 33*(2), 297–298.

Marmot, M. (2004b). *The status syndrome: How social standing affects our health and longevity*. New York: Henry Holt.

Marmot, M., & Wilkinson, R. G. (Eds.). (2006). *Social determinants of health* (2nd ed.). Oxford: Oxford University Press.

Morrow, D., Clark, D., Tu, W., & Wu, J. (2006). Correlates of health literacy in patients with chronic heart failure. *The Gerontologist, 46*(5), 669–676.

Myers, G. C., & Manton, K. G. (1984). Compression of mortality: Myth or reality. *The Gerontologist, 24*(4), 346–353.

National Research Council. (2001). *Preparing for an aging world: The case for cross-national research*. Washington, DC: National Academy Press.

OECD. (2010a). *Factbook 2010: Economic, environmental and social statistics*. Paris: Organisation for Economic Co-operation and Development.

OECD. (2010b). *Health data 2010, selected data, risk factors*. http://www.oecd.org/document/16/0,3343,en_2649_34631_2085200_1_1_1_1,00.html. Accessed Sept 2010.

Oeppen, J., & Vaupel, J. W. (2002). Broken limits to life expectancy. *Science, 295*(5570), 1029–1031.

Olshansky, S. J., & Ault, A. B. (1986). The fourth stage of the epidemiologic transition: The age of delayed degenerative disease. *The Milbank Memorial Fund Quarterly, 64*(3), 355–391.

Population Reference Bureau. (2010). *2010 World population data sheet*. Washington, DC: Population Reference Bureau.

Robine, J., Cheung, S. L. K., Horiuchi, S., & Thatcher, A. R. (2008a). Is there a limit to the compression of mortality? In *Living to 100 and beyond*. Orlando, USA: Society of Actuaries. http://www.soa.org/library/monographs/retirement-systems/living-to-100-and-beyond/2008/january/mono-li08-03-cheung.pdf. Accessed Feb 2011.

Robine, J., Cheung, S. L. K., Horiuchi, S., & Thatcher, A. R. (2008b). Is the compression of morbidity a universal phenomenon? In *Living to 100 and beyond*. Orlando, USA: Society of Actuaries. http://www.soa.org/library/monographs/retirement-systems/living-to-100-and-beyond/2008/january/mono-li08-04-cheung.pdf. Accessed Feb 2011.

Rogers, R. G., & Hackenberg, R. (1987). Extending epidemiologic transition theory: A new stage. *Social Biology, 34*, 234–243.

Rowe, J. W., & Kahn, R. L. (1998). *Successful aging* (large print ed.). New York: Random House.

Sadowski, C. A. (2010). Providing health information to older adults. *Reviews in Clinical Gerontology, 21*(1), 55–66.

Senate Community Affairs Committee. (2005). *Quality and equity in aged care*. Canberra: Commonwealth of Australia. http://www.aph.gov.au/senate/committee/clac_ctte/completed_inquiries/2004-07/aged_care04/report/index.htm. Accessed June 2011.

Shanas, E., Townsend, P., Wedderburn, D., Friis, H., Mihoj, P., & Stehouwer, J. (Eds.). (1968). *Old people in three industrial societies*. London: Routledge and Kegan Paul.

Stampfer, M. (2004). New insights from the British doctors study. *British Medical Journal, 328*, 1507.

Suzuki, M., Willcox, B., & Willcox, C. (1995). *Centenarians in Japan*. Tokyo: Nakayamashoten. See: Okinawa centenarian study. http://www.okicent.org/study.html. Accessed June 2011.

Temple, J. B. (2006). Household factors associated with older Australian's purchasing a varied diet: Results from household expenditure data. *Nutrition and Dietetics, 63*(1), 28–35.

Volker, D., & Caterson, I. (2001). Nutrition and senescence: Healthy ageing. *Australasian Journal on Ageing, 20*, 114–123.

Wadsworth, M. (1999). Early life. In M. Marmot & R.G. Wilkinson (Eds.), *Social determinants of health* (pp. 44–63). Oxford: Oxford University Press.

WHO (World Health Organisation). (1998). *Health promotion glossary*. WHO/HPR/HEP/98.1 Geneva: WHO. http://www.who.int/hpr/NPH/docs/hp_glossary_en.pdf. Accessed Jan 2011.

WHO (World Health Organisation). (2009a). *Global health risks: Mortality and burden of disease attributable to selected major risks*. Geneva: World Health Organisation Press.

WHO (World Health Organisation). (2009b). *WHO Statistical information system*, http://www.who.int/whosis/data/Search.jsp; and WHO *Core health indicators*, http://apps.who.int/whosis/database/core/core_select.cfm. Accessed Apr 2009.

Wilkinson, R. G. (1999). Putting the picture together: Prosperity, redistribution, health and welfare. In M. Marmot & R. G. Wilkinson (Eds.), *Social determinants of health* (pp. 256–274). Oxford: Oxford University Press.

Willcox, B. J., Yano, K., Chen, R., et al. (2004). How much should we eat? The association between energy intake and mortality in a 36-year follow-up study of Japanese-American men. *The Journals of Gerontology: Biological Sciences, 59A*(8), 789–795.

Willcox, D. C., Willcox, B. J., He, Q., Wang, N., & Suzuki, M. (2008). They really are that old: A validation study of centenarian prevalence in Okinawa. *The Journals of Gerontology: Biological Sciences, 63A*(4), 338–349.

Wilmoth, J. R., & Horiuchi, S. (1999). Rectangularization revisited: Variability of age at death within human populations. *Demography, 36*(4), 475–495.

Chapter 5
Survival and Health

To add life to years, not just years to life (From the cover of the first issue of the Journal of Gerontology, January 1946).

5.1 Population Health

Ill health is personally, socially and economically among the most costly concerns in aging populations. Although it is widely held that ongoing improvements are possible, by no means all current health changes in aging societies are positive. This chapter compares societies in terms of indicators of population health, namely life expectancy, expectations of healthy life, the changing prevalence of disabilities and the main causes of lost years of healthy life. To obtain an overview of health trends the chapter draws on some of the World Health Organisation's (WHO) innovations in the measurement of population health. Average life expectancy at birth has long been one of the most readily available and most commonly employed indicators of population health. Its main disadvantage is that it is only an indirect indicator: it provides no specific information about the duration of life lived without impairments. In response to the need for direct measures of the duration of healthy life, the WHO developed a new data set that combined information on survival and health status into a new measure of overall population health, termed 'healthy life expectancy'. It was based on current rates of ill-health and mortality. Besides this the WHO has also developed measures of 'the burden of disease' which reveal the impact on health of particular diseases and injuries. Although the new measures are partly experimental, often relying on extensive estimation, they help to address key concerns in aging populations.

D.T. Rowland, *Population Aging: The Transformation of Societies*, International Perspectives on Aging 3, DOI 10.1007/978-94-007-4050-1_5,

5.2 Life Expectancy at Older Ages

Late in the epidemiologic transition life expectancy at age 65 increases more rapidly than before, as illustrated in Fig. 5.1. Female life expectancies at birth are shown on the horizontal axis, with corresponding figures for males underneath – male life expectancy at birth is typically 3–6 years lower. The graph is based on model data representing averages for Western countries through time. Conspicuous are the large increases in life expectancy at 65 associated with the lowest levels of mortality. Thus in a cohort of females, when life expectancy at birth is 85, life expectancy at 65 is just over 22 years. Similarly, for a male cohort with a life expectancy at birth of 79, the survivors have a life expectancy of 18 years at age 65.

Because of the selectivity of survival those who reach age 65 are not a representative cross-section of people born 65 years earlier: they have better than the average length of life for the cohort as a whole. As discussed in Chap. 4, there is a greater likelihood of early death among the socially and economically disadvantaged, those least averse to taking risks with their health, such as through smoking, together with people engaged in hazardous occupations. Selectivity in survival also occurs because of the effects of genetics. Some argue that genetically-related diseases take their greatest toll in the first decades of life, with the result that those who live longer are, genetically, a more select group (Rowe and Kahn 1998). Assuming that genetic influences are similar in all populations, a substantial part of the differences between national populations in life expectancy at all ages reflects the 'social component' of health – the aspects of health potentially subject to modification through decisions and policies.

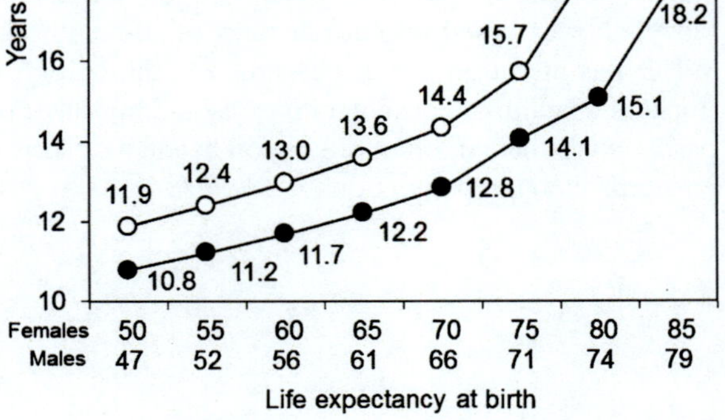

Fig. 5.1 Life expectancy of females and males at age 65 (Sources: West Model Life Tables: Coale and Demeny 1983; Coale and Guo 1990)

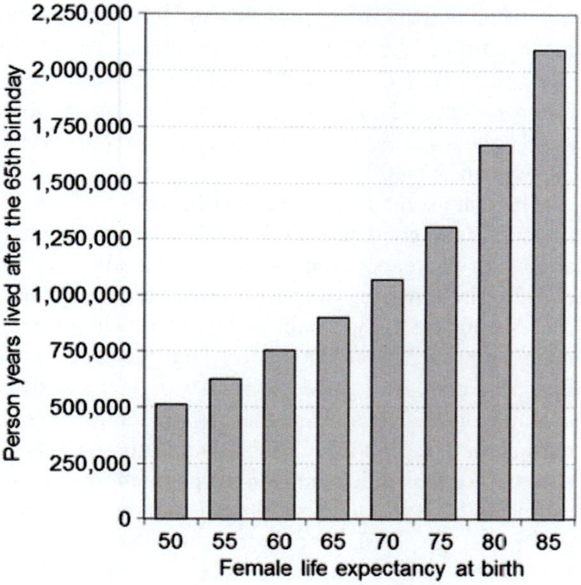

Fig. 5.2 Changes in person years lived from the 65th birthday by a cohort of 100,000 females (*Note*: Person years are the sum of all the years members of a population live during a specified interval) (Source: see Fig. 5.1)

Associated with the extension of life expectancy at age 65 is a great increase is the average number of years lived from that age. This is measured by changes in the person years lived, by a cohort numbering 100,000 at birth, from the survivors' 65th birthdays to the end of their lives (Fig. 5.2). When life expectancy is 70 years, the surviving members of the cohort live approximately 1.1 million person years after age 65. This figure almost doubles when life expectancy reaches 85 years: the survivors then live 2.1 million person years after age 65. Such increases in person years lived have major implications for social security systems because they denote a huge expansion in the period of time during which people may be receiving age pensions. Raising pensionable ages, means testing and encouraging private saving for retirement are responses to this relatively new phenomenon.

5.3 Healthy Life Expectancy

In aging populations, a significant measure of progress in adding life to years is the expectation of healthy life, or health-adjusted life expectancy (HALE). This refers to the expected number of years lived without reduced functioning due to ill health at any age. The calculations incorporate the sum of all episodes of ill health throughout life and use statistics on various health states to weight data on person years lived by populations. A weight or multiplier of 1 applies to years of good health, while weights of less than 1 denote years in ill health – the lower the weight the greater the severity of impairment (Mathers 2002: 181). The WHO first published figures on healthy life expectancy in 1999. The ensuing series is the only source of information on health expectancies that can be compared across countries (ibid.: 177 and 199). Self-reported health data, for example, tend to be unreliable sources for cross-national comparisons because groups or populations that are wealthier and have greater access to resources tend to report worse health (Sadana et al. 2002: 383; Johansson 2003: 481).

There are some marked differences between countries in their attainment of long life and life in good health (Table 5.1). Japanese females lead with a life expectancy at birth of 86 years (2006–2010), and a healthy life expectancy at birth of 78 years (2003). These figures were 2 or 3 years ahead of those for the closest ranking countries, including Italy, Spain and France. The corresponding figures for Japanese males, however, were equaled or nearly equaled in several Northern and Western European countries, together with Australia. Latitude for improvement is evident, for example, in Australia where there is still a high prevalence of lifestyle-related health risks as well as pronounced social inequality (AIHW 2008: 131ff). The latter is most extreme for Australia's indigenous population in which males in 1996–2001 had an estimated life expectancy at birth of 59 years and females 65 years (ibid.: 68).

On average, in OECD countries, females live about 8 years of their total lifetimes in ill health while the figure for men is 6 years. Women have a greater prevalence of disabilities, more reported symptoms and more non-life-threatening diseases. This is attributed to greater female longevity, as well as to gender differences in health-related behaviour and use of health services (Christensen et al. 2009: 1204). There is a considerable margin for potential change in the United States where females trail more than 6 years behind Japan in life expectancy and 7 years behind in healthy life expectancy. Females in the United Kingdom and Denmark also have relatively low expectancies of total life and healthy life (Table 5.1).

As in some accounts of why the USA's birth rate has been close to replacement level (see Chap. 8), social inequality is a presumed reason for its relatively high mortality. Many in the United States have low incomes, which leaves them at greater risk of disadvantage in access to education, good nutrition, and medical services (Munnell et al. 2004). The impact of unhealthy lifestyles is further evident in the American obesity epidemic. The rising prevalence of obesity is a relatively new issue which has the potential to block improvements in life expectancy and healthy life expectancy in many societies (see Olshansky et al. 2005).

Some parts of Eastern Europe have had the poorest health outcomes of all the 'developed' countries. In the Russian Federation, for example, male life expectancy at birth and healthy life expectancy were both 19 years behind Japan's, despite the country having relatively high levels of education (National Center for Education Statistics 2007: iii–v). The World Bank classified the Russian Federation's economy in 2004 as 'lower middle income', the income rank immediately above that of India and Bangladesh (Table 5.1). With the fall of communism and the break-up of the Soviet Union, life expectancy in Eastern Europe declined during the late 1980s and through the 1990s (Shkolnikov et al. 2001: 917; Men et al. 2003). This reflected more widespread poverty, the need for longer hours of work to earn sufficient income, and falls in production, real wages and living standards. One estimate

Table 5.1 Life expectancy and healthy life expectancy at birth and age 60, selected countries

Country	World Bank economy classification[a]	Life expectancy at birth		Healthy life expectancy at birth		Life expectancy at 60		Healthy life expectancy at 60	
Sex:		M	F	M	F	M	F	M	F
Year:	2004	2006	2006	2003	2003	2000	2000	2002	2002
North America and Australasia									
United States	High income	75	80	67	71	19.6	23.1	15.3	17.9
Canada	High income	78	83	70	74	19.8	23.9	16.1	19.3
Australia	High income	79	84	71	74	20.7	24.6	16.9	19.5
New Zealand	High income	78	82	69	72	20.3	24.0	16.0	18.2
Northern and Western Europe									
United Kingdom	High income	77	81	69	72	18.8	22.7	15.7	18.1
France	High income	77	84	69	75	20.8	26.0	16.6	20.4
Switzerland	High income	79	84	71	75	20.6	25.2	17.1	20.4
Belgium	High income	77	82	69	73	19.5	23.9	15.7	19.1
Netherlands	High income	78	82	70	73	18.9	23.7	15.5	18.4
Denmark	High income	76	81	69	71	18.8	21.6	15.2	17.2
Norway	High income	78	83	70	74	19.6	23.9	16.2	18.9
Sweden	High income	79	83	72	75	20.6	24.3	17.1	19.6
Finland	High income	76	83	69	74	18.7	23.5	15.7	18.9
Germany	High income	77	82	70	74	18.7	23.2	15.9	19.0
Austria	High income	77	83	69	74	19.4	23.9	16.2	19.3
Southern Europe									
Italy	High income	78	84	71	75	20.3	24.9	16.4	19.4
Greece	High income	77	82	69	73	19.7	23.0	16.0	18.1
Slovenia	High income	74	82	67	72	17.5	22.4	14.3	18.1
Spain	High income	78	84	70	75	20.0	24.7	16.4	19.9
Portugal	High income	75	82	67	72	18.0	22.3	14.9	17.7

(continued)

Table 5.1 (continued)

Country	World Bank economy classification[a]	Life expectancy at birth		Healthy life expectancy at birth		Life expectancy at 60		Healthy life expectancy at 60	
Sex:		M	F	M	F	M	F	M	F
Year:	2004	2006	2006	2003	2003	2000	2000	2002	2002
Eastern Europe									
Poland	Upper-middle income	71	80	63	68	16.8	21.5	12.8	16.1
Czech Republic	Upper-middle income	73	80	66	71	17.0	21.2	13.5	16.8
Croatia	Upper-middle income	72	79	64	69	15.5	20.7	12.5	16.1
Bulgaria	Lower-middle income	69	76	63	67	15.5	19.6	12.5	15.0
Russian Fed.	Lower-middle income	60	73	53	64	13.3	18.3	9.7	14.0
Ukraine	Low-income	61	73	55	64	14.0	18.5	10.3	13.7
Hungary	Upper-middle income	69	78	62	68	14.9	19.6	12.1	16.0
Latvia	Lower-middle income	65	76	58	68	15.8	20.8	11.3	15.7
Asia and Latin America									
Japan	High income	79	86	72	78	21.3	26.8	17.5	21.7
Singapore	High income	78	83	69	71	19.5	22.7	14.5	16.3
China	Lower-middle income	72	75	63	65	16.6	20.4	13.1	14.7
Indonesia	Low-income	66	69	57	59	15.5	17.5	10.7	11.5
India	Low-income	62	64	53	54	14.6	17.7	10.8	11.4
Bangladesh	Low-income	63	63	55	53	14.7	15.7	11.1	11.1
Pakistan	Low-income	62	63	54	52	15.6	17.4	11.4	11.4
Mexico	Upper-middle income	72	77	63	68	19.7	21.7	14.4	16.2
Brazil	Upper-middle income	68	75	57	62	16.2	19.6	11.6	13.7

Sources: WHO (2004a, b), World Bank (2004)

[a] The World Bank divided countries into income groups according to 2004 gross national income (GNI) per capita. The groups are low income (US$825 or less), lower middle income (US$826–3,255), upper middle income (US$3256–10,065), and high income (US$10,066 or more)

indicated that more than a third of the Russian population was living in poverty in the 1990s (Shaw et al. 2006: 206). Around the mid-1990s cardiovascular diseases, especially coronary heart disease, accounted for more than 50% of the difference in mortality between Eastern and Western Europe while external causes, including accidents and homicide, accounted for another 20% (Marmot and Wilkinson 1999: 7 and 106). Further factors in the rise in ill health and mortality were alcohol-related deaths of men, a decline in the quality of medical care, air pollution, smoking, poor diet and lack of exercise. In the Russian Federation, alcohol abuse has led to higher adult male mortality as well as to disability from accidents, violence and cardiovascular diseases (Lopez et al. 2006: 87–88). Despite some improvements in survival, mainly due to reductions in deaths from alcohol-related causes (Shkolnikov et al. 2001: 917), the Eastern European disadvantage in survival has persisted.

International data for comparing life expectancy with healthy life expectancy at older ages are mostly available for age 60 instead of age 65 and the lower age limit is therefore employed here. In OECD countries men aged 60 spend 3.5 years of their remaining life in ill health while the figure for women is 4.7 years. This pattern also holds for a range of different national life expectancies (Table 5.1). Quite long duration of ill health in later life is therefore an established feature. The figures tend to be a little higher in the developing countries, especially for older women. In OECD countries, the duration of ill health amounts to approximately 20% of life from age 60. However, in the Russian Federation the figures are 24% for older women and 27% for older men, which parallels the poorer health outcomes for Eastern European populations generally. Despite this, the number of years that older people in the Russian Federation spend in ill health is no higher than the OECD averages. This probably signifies earlier onset of life-threatening conditions and poorer chances of survival from them (Andreev et al. 2003).

The WHO data on healthy life expectancy refer to just 1 year rather than trends through time – which are an important consideration because of the need for accurate forecasts of health care costs (Freedman et al. 2004: 417). Information for Great Britain in the period 1981–2003, although not very comparable with the WHO figures, imply that the difference between life expectancy at 65 and healthy life expectancy at 65 remained about 5 or 6 years for females and 3 or 4 years for males: there was no clear trend towards a widening or narrowing of the gap. In Great Britain in 2003, 65 year old females had a life expectancy of 19.3 years and a healthy life expectancy of 14.3 years, while for males the figures were 16.4 and 12.3 years (National Statistics 2007).

Health data pose many problems in comparisons through time, but there is accumulating evidence from the United States of improvements in health in the young-old ages (65–74 years). One study sought to resolve inconsistencies in national surveys of activity limitations among persons aged 70. It determined that there was good evidence of an appreciable decline in the proportions of the community-based population reporting difficulty with daily activities during the 1990s (Freedman et al. 2004: 435). Another American study found evidence of improvement for people aged 65–74 years, among whom there was a marked reduction over the 30 years to 1990 in the prevalence of three important precursors to chronic disease: high blood pressure, high cholesterol levels and smoking. The authors also reported reductions in the prevalence of arthritis, arteriosclerosis, dementia, stroke and emphysema (Rowe and Kahn 1998: 17–18). In the United States there has also been a consistent decline in the proportions experiencing disabilities in later life, such as difficulty walking, paralysis, deafness and blindness (Costa 2002 cited by Klein 2004: 222). Improvements in vision owe much to advances in cataract surgery, which has become the most frequently performed surgical procedure in developed countries (Christensen et al. 2009: 1199). Overall, disability rates in the United States declined in the 23 years to 2005, with reduced prevalence of heart and circulatory conditions and vision impairments playing a major role (Kinsella and He 2009: 53).

As in the United States, in OECD countries the overall prevalence of age-specific chronic disability has been declining since the early 1990s despite population aging (Spillman 2004, cited by Howse 2005: 4). An extensive review, published in *The Lancet*, reported a trend in developed countries towards falling rates of disabilities in activities of daily living since the 1980s (Christensen et al. 2009: 1200 and 1203). It also reported evidence of a decrease in the number of years lived with the most severe levels of disability (Robine et al. 2003, cited by Christensen et al. 2009: 1201). Another main finding was that previously 'silent' diseases, such as type 2 diabetes and some cancers, are now diagnosed earlier and receive better treatment. This extends the period of morbidity but with improved function (ibid.: 1198). The review concluded that:

> Most evidence for people aged younger than 85 years suggests postponement of limitations and disabilities, despite an increase in chronic diseases and conditions. This apparent contradiction is at least partly accounted for by early diagnosis, improved treatment, and amelioration of prevalent diseases so that they are less disabling. … For people aged older than 85 years, the situation is less clear. … Individuals who survive longest have a health profile that is, in many respects, similar to that of individuals who are a decade or so younger. (ibid. 2009: 1204–5).

Despite this positive perspective, other research has not confirmed a general trend towards a decline in the prevalence of severe disabilities in all OECD countries. A review of ADL (activities of daily living) disability rates at ages 65 and over in 12 OECD countries during the 1990s found clear evidence of a decline in severe disability among elderly people in only five of the countries; three countries had an increasing rate and two a stable rate. The prevalence of ADL disability also differed considerably between countries (Lafortune et al. 2007, cited by Robine et al. 2008). A reason why this study seems to contradict others cited above is its focus on severe disabilities, which have their highest prevalence at the oldest ages. Some of the other studies focused on the young-old rather than the oldest-old. In developed countries, most people experiencing episodes of ill health in their 60s and 70s will not be severely disabled and will remain capable of living independently much of the time. Early diagnosis and effective interventions enable many to live with disease-related disabilities moderated or eliminated.

Length of life with disabilities and ill health is one of the main issues that will affect the economic sustainability of population aging. A key question is whether current developments are narrowing the gap between life expectancy and healthy life expectancy. Research on the United States population found that active life expectancy at age 65 (years of life spent in a healthy or nondisabled state) increased by 5 years between 1935 and 1999, while life expectancy increased by 6 years. The authors' projections suggested that active life expectancy at age 65 would increase from 13.9 years in 1999 to 16.4 years in 2022, with further increases to follow. They concluded that extension of life in "a socially or economically productive state" would enable continuing increases in retirement ages – to 72 years in 2022. Such a strategy could keep constant, at the 1935 level, the length of time active retired people received social security benefits (Manton et al. 2006: 94–98). Longitudinal data for different countries are needed to examine this question further: trends in disabilities and diseases differ through time within and between societies and there are changes working in opposite directions:

> Elderly people are reporting higher morbidity levels in almost all repeated cross-sectional surveys in developed countries. From year to year, they are more educated, which possibly entails a better medical awareness and literacy. They see doctors more often. They may have a higher level of health expectation and lower threshold before going to see the doctor or reporting health problems. They may also have healthier life styles. On the other hand doctors diagnose diseases earlier and are more efficient in treating them, slowing down the morbidity processes, preventing complications and postponing deaths. (Robine et al. 2008: 3).

5.4 The Burden of Disease

Statistics on causes of death have long served as indicators of population health, but they conceal much of the human costs of shortened lives and years of illness and incapacity. Similarly, statistics on the incidence and prevalence of diseases, now commonly available in developed countries, record only the number of new or existing cases rather than the overall impact of diseases and disabilities on quality of life. Another innovation in health statistics, emerging during the 1990s, through research sponsored by the World Bank and the World Health Organization, has been the development of statistics on 'the burden of disease', which seek to address these gaps in conventional health data. This section is based on some of the measures of the burden of disease at ages 60 and over.

The best-known causes of disabilities in later life are those that become more prevalent with age, notably dementia, arthritis, osteoporosis and fractures – especially hip fractures due to falls. Yet many other conditions are leading causes of the burden of disease at older ages, because they are responsible for years of illness as well as the loss of potential years of life. Information on this is available from the extensive estimates in the publication: *Global Burden of Disease and Risk Factors* (Lopez et al. 2006). The data from this source, in Tables 5.2 and 5.3, refer to countries in the World Bank's 'high income countries' – predominantly countries in Europe (excluding Eastern Europe), North America and Australasia, together with Japan, Singapore, South Korea and parts of the Middle East and the Caribbean (ibid.: 94). The measures of the burden of disease in the Tables are DALYs, or 'disability adjusted life years', which were also used in Chap. 4 to evaluate consequences of health risks. One DALY is one lost year of healthy life. DALYs for a disease or health condition are the sum of the years of healthy life

Table 5.2 Lost years of healthy life (DALYs) by age, 2001

Region	0–14	15–59	60+	Total
Percentage of total lost years of healthy life				
High income countries	6.8	42.2	51.0	100.0
Low and middle income countries	36.6	42.6	20.8	100.0
World	33.7	42.6	23.7	100.0
Lost years of healthy life per 1,000 population				
High income countries	59	109	427	161
Low and middle income countries	306	189	656	266
World	283	177	590	250

Source: Calculated from Lopez et al. (2006: 180–181, 222–223, 228–229)
Note: A detailed definition of DALYs is given in Lopez et al. (2006: 3)

Table 5.3 Leading causes of lost years of healthy life (DALYs) at ages 60 and over, high income countries, 2001

	60–69	70–79	80+	Total	DALYs (000s)
Population (millions)	84	63	31	178	
Total DALYs all causes (thousands)	25,692	29,539	20,777	76,008	
Lost years of healthy life per 1,000 population	306	469	670	427	
Cause	Percentage of lost years of healthy life				
Cardiovascular diseases	7.9	11.7	10.3	29.8	22,679
Ischemic heart disease	3.6	5.0	3.8	12.4	9,446
Cerebrovascular disease	2.5	3.8	3.1	9.5	7,209
Malignant neoplasms	9.3	9.1	3.7	22.0	16,758
Trachea, bronchus, and lung cancers	2.3	2.1	0.6	5.0	3,766
Colon and rectal cancers	1.2	1.2	0.6	3.0	2,260
Neuropsychiatric conditions	3.4	4.9	5.3	13.7	10,413
Alzheimer's and other dementias	1.4	3.5	4.5	9.4	7,135
Respiratory diseases	2.2	2.6	1.7	6.5	4,930
Chronic obstructive pulmonary disease	1.4	1.7	0.9	4.0	3,046
Sense organ diseases	2.9	2.1	0.5	5.5	4,179
Hearing loss, adult onset	2.1	1.5	0.3	3.8	2,926
Musculoskeletal diseases	1.9	1.8	0.7	4.5	3,385
Osteoarthritis	1.4	1.3	0.4	3.1	2,325
Digestive diseases	1.5	1.5	1.1	4.1	3,140
Diabetes mellitus	1.2	1.2	0.6	3.1	2,340
Respiratory infections	0.4	0.9	1.3	2.5	1,899
Unintentional injuries	0.8	0.8	0.6	2.2	1,642
Total	31.6	36.6	25.7	93.9	71,365

Source: Calculated from Lopez et al. (2006: 222–227)

lost due (a) to years lived in less than full health and (b) to premature mortality measured with reference to a global standard life expectancy, based on the highest observed figures (ibid.: 3 and 47–50). Thus DALYs account for the difference between healthy life expectancy and the potentially attainable life expectancy. To take appropriate account of sex differentials in mortality, the WHO has used an evidence-based intrinsic biological difference in male and female life expectancies of 2.5 years – much less than the gap of up to 7 years observed in developed countries (ibid.: 48).

Calculation of the DALYs entailed the use of weighting systems, for instance to allow for variations in the severity of diseases. The measurements involve extensive estimation but serve as an essential baseline for further refinements. In 2001, the estimated global average burden of disease for all ages was 250 lost years of healthy life/DALYs per 1,000 population. Almost two thirds of this was due to premature deaths (ibid.: 87). Deaths of children in developing countries contribute substantially to the high figure, partly because such countries have younger age structures and partly because every child's death results in decades of potential life lost. In 2001, children under 15 accounted for 34% of the burden of disease in the world as a whole, compared with 7% of the burden in high income countries (Table 5.2).

As populations grow older, in conjunction with socio-economic development and longer life expectancies, the lost years of healthy life become concentrated in the older ages. Thus an estimated 51% of the total lost years of healthy life occurred at ages 60 and over in high income countries, compared with about 21% in the low and middle income countries The rates at which years of healthy life are lost are very much higher at all ages in the populations of low and middle income countries (Table 5.2). It is sometimes thought that aging adds to health care costs because these are greatest in the last year of life. However, Healy (2004: vii) argued that health costs for the elderly in the last year of life are likely to be lower than those of younger patients, who more often receive more extensive and expensive treatments.

For high income countries, Table 5.3 shows the main groups of causes of lost years of healthy life together with the major diseases within them. Cardiovascular diseases are by far the leading causes of ill health and premature mortality at ages 60 and over, accounting for 30% of the total lost years of healthy life. Ischemic heart disease and cerebrovascular disease are the principal diseases within this group. For both diseases the known and potentially controllable risk factors are high blood pressure, high cholesterol, smoking, overweight and obesity, physical inactivity and low fruit and vegetable intake (Lopez et al. 2006: 264). Malignant neoplasms rank second and are responsible for 22% of lost years of healthy life. Cancers of the respiratory system and bowel account for 36% of this total. The third most dominant group of causes is neuropsychiatric conditions, within which Alzheimer's disease and other dementias take the greatest toll, at a level similar to that of cerebrovascular disease. Overall, the three leading groups of causes account for two thirds of the total lost years of healthy life at ages 60 and over.

The other causes listed in Table 5.3 – such as respiratory diseases, diabetes, and osteoarthritis – account for relatively small shares of the burden of disease.

Nevertheless, because the three main groups of diseases take such a high toll, a small share of a very large total burden of disease is significant in itself. Moreover, many diseases and disabilities have major consequences not only for patients but also for their families and support networks. Such impacts can be prolonged and severe, for instance, where a disease impairs an individual's mental functioning, mobility or self-care. Falls – and resulting fractures, concussion and other injuries – do not figure among the leading causes in Table 5.3 because they accounted for only 0.8% of the total lost years of healthy life. Nevertheless they are an important trigger for high levels of dependency for many older people. Other significant causes of disabilities are similarly overshadowed in any broad summary in which the main chronic diseases dominate. For example, incontinence is a major social and medical issue, one which is enormously costly to governments, communities and individuals (Department of Health and Ageing 2008: 23). Potentially, incontinence can be corrected or ameliorated in many cases, but it remains a major cause of the institutionalization of the elderly. More effective prevention, treatment and management of incontinence offers great potential for restoring people's dignity and independence, reducing nursing time spent on this problem and avoiding premature admission to an aged care institution. Also prominent in public consciousness of negative consequences of aging are depression (unipolar depressive disorders) and Parkinson's disease, despite their accounting for only 1.4% and 1.2% respectively of the lost years of healthy life at older ages in high income countries. One of the discoveries of burden of disease research has been the importance of depression as a cause of ill health. For all age groups in high income countries unipolar depressive disorders were responsible for 5.6% of the lost years of healthy life in 2001 (Lopez et al. 2006: 89). Nevertheless the prevalence of depressive disorders declines with age: it is lowest at older ages in the United States and Australia. It appears that depression and anxiety, which commonly co-occur, become less prevalent with age, for instance through learned coping strategies and less emotional responses.

Among the most intractable of disabling diseases, and most costly in terms of demands on families, nursing staff and health care systems, are Alzheimer's disease and other forms of dementia. Their share of the burden of disease in later life increases with advancing age. At ages 80 and over in high income countries dementia accounts for 17% of the lost years of healthy life compared with 9% at ages 70–79. Alzheimer's disease is most common, followed by vascular dementia. In Australia they account for about 50% and 20% of cases respectively (Access Economics 2009: 3). Hopes for reducing the numbers of cases currently resides particularly in addressing the risk factors for vascular dementia which overlap with those for ischemic heart disease and stroke (Rowe and Kahn 1998: 130; Alagiakrishnan 2007). Recommended strategies for individuals to reduce the risk of dementia, or delay its onset, include being physically, mentally and socially active, maintaining a healthy diet and weight, avoiding smoking and immoderate alcohol consumption, having regular health checks on blood pressure, cholesterol and blood sugar, and avoiding falls and other causes of head trauma (Alzheimer's Australia 2011).

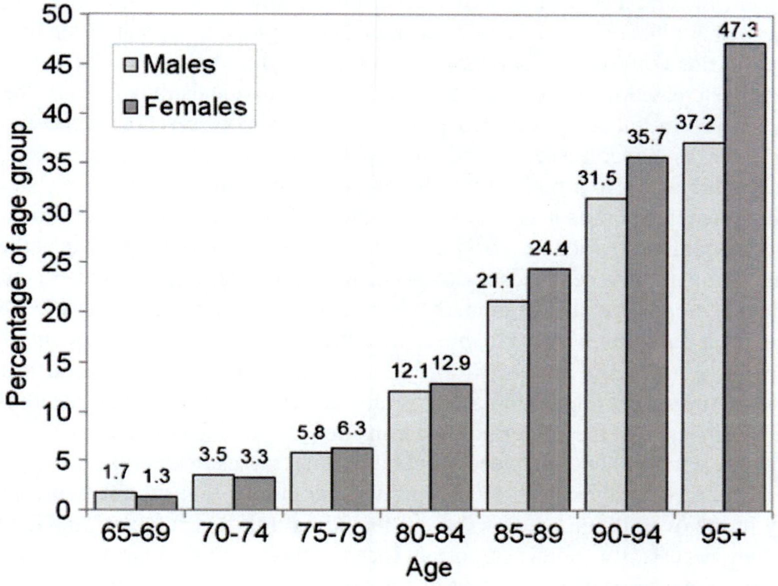

Fig. 5.3 Estimated dementia prevalence rates, 2009 (Source: Access Economics 2010: 12)

Figure 5.3 further illustrates the steep rise in the prevalence of dementia with age; the estimated rates derive from published Australian and international epidemiological studies and meta-analyses (Access Economics 2010: 17). Rates of this magnitude, in conjunction with numerical growth at older ages in virtually all national populations, portend a 'pandemic' of dementia in coming decades. Applying the rates to the UN's medium variant projections for Europe produces a prevalence of about 12 million cases aged 65 and over in 2025 and 19 million in 2050. Over the second quarter of the century the percentage of the aged population affected would increase from 8% to 10%. Much higher numbers of cases are projected for Africa, Asia and Latin America (see Chap. 15). However, because of differences between the methodologies employed in various studies to obtain data on prevalences, true rates of dementia by age and sex remain indeterminate (AIHW 2007: 54).

5.5 Conclusion

Long life is an achieved goal for many countries and its limits have yet to be reached. Yet even in the most prosperous countries, healthy life expectancy at age 60 is well below overall life expectancy. Also, in the same countries, the difference between male and female life expectancies at birth is probably double that due solely to biological factors. The nature of the potentially preventable components of these differences attains increasing economic and social significance as growing numbers and proportions of people reach their seventh and eighth decades of life.

This has created major research interest in how individuals may achieve 'healthy aging' or 'successful aging' (see Chap. 10). Associated concerns are the need to contain growth in government expenditure on health services and pharmaceuticals. Demographic aging is a major force for increases in health expenditure, but compounding factors are rises in hospitalization, more frequent consultations with doctors and other health professionals, and greater use of high technology for medical diagnosis and treatment. In recent decades scientific innovation has made a greater contribution than demographic change to increases in health care spending (Howse 2006: 5).

Cardiovascular diseases and cancers account for the greater part of the burden of disease in later life and are necessarily a major focus for current medical research and expenditure. Yet population aging is altering disease patterns and health issues in the fourth stage of the epidemiologic transition. Neurodegenerative diseases have emerged as a major concern, closely linked with increases in the numbers of people reaching advanced ages. Non-fatal disabling conditions that do not figure prominently in the overall burden of disease – including arthritis, fractures and osteoporosis – are necessarily another important focus for prevention and treatment in aging populations because of their potential for causing severe disabilities. A challenge for the future will be to increase the effectiveness of preventive strategies, as well as early detection and treatment of diseases, without raising the negative consequences for quality of life envisaged in Barsky's paradox.

References

Access Economics. (2009). *Keeping dementia front of mind: Incidence and prevalence 2009–2050*. Canberra: Alzheimer's Australia.

Access Economics. (2010). *Caring places: Planning for aged care and dementia 2010–2050*. Canberra: Alzheimer's Australia.

AIHW (Australian Institute of Health and Welfare). (2007). *Dementia in Australia: National data analysis and development* (AIHW Cat. No. AGE 53). Canberra: Australian Institute of Health and Welfare.

AIHW. (2008). *Australia's Health 2008*. (AIHW Cat. No. AUS 99). Canberra: Australian Institute of Health and Welfare.

Alagiakrishnan, K. (2007). *Vascular dementia, eMedicine, World Medical Library*. http://www.emedicine.com/m ed/topic3150.htm, Article Last Updated: Aug 27, 2007. Accessed Sept 2007.

Alzheimer's Australia. (2011). *Mind your mind*. http://www.alzheimers.org.au/min-your-mind.aspx. Accessed Feb 2011.

Andreev, E. M., McKee, M., & Shkolnikov, V. M. (2003). Health expectancy in the Russian Federation: A new perspective on the health divide in Europe. *Bulletin of the World Health Organization, 81*(11), 778–787.

Christensen, K., Doblhammer, G., Rau, R., & Vaupel, J. W. (2009). Ageing populations: The challenges ahead. *The Lancet, 374*, 1196–1208.

Coale, A. J., & Demeny, P. (1983). *Regional model life tables and stable populations*. New York: Academic Press.

Coale, A. J., & Guo, G. (1990). New regional model life tables at high expectation of life. *Population Index, 56*, 26–41.

Costa, D. L. (2002). Changing chronic disease rates and long term decline in functional limitation among elderly men. *Demography, 39*(1), 119–137.

Department of Health and Ageing. (2008). *Ageing and aged care in Australia*. Canberra: Australian Government, Department of Health and Ageing.

Freedman, V. A., et al. (2004). Resolving trends in old-age disability: Report from a technical working group. *Demography, 41*(3), 417–441.

Healy, J. (2004). *The benefits of an ageing population* (The Australia Institute Discussion Paper, No. 63). Canberra: The Australia Institute.

Howse, K. (2005). Policies for healthy ageing. *Ageing Horizons*, Issue 2: 3–15.

Howse, K. (2006). *Increasing life expectancy and the compression of morbidity: A critical review of the debate* (Oxford Institute of Ageing Working Paper No. 206). Oxford: Oxford Institute of Ageing.

Johansson, S. R. (2003). Health transition. In P. Demeny & G. McNicoll (Eds.), *Encyclopedia of population* (pp. 479–483). New York: Macmillan Reference.

Kinsella, K., & He, W. (2009). *An aging world 2008* (U.S. Census Bureau, International Population Reports, P95/09-1). Washington, DC: U.S. Government Printing Office.

Klein, H. S. (2004). *A population history of the United States*. Cambridge: Cambridge University Press.

Lafortune, G., Balestat, G., & The Disability Study Expert Group Members. (2007). *Trends in severe disability among elderly people: Assessing the evidence in 12 OECD countries and future implications* (OECD Health Working Papers, No. 26). Paris: OECD.

Lopez, A. D., Mathers, C. D., Ezzati, M., Jamison, D. T., & Murray, C. J. L. (2006). *Global burden of disease and risk factors*. Washington, DC: Oxford University Press and the World Bank.

Manton, K. G., Gu, X., & Lamb, V. L. (2006). Long-term trends in life expectancy and active life expectancy in the United States. *Population and Development Review, 32*(1), 81–105.

Marmot, M., & Wilkinson, R. G. (Eds.). (1999). *Social determinants of health*. Oxford: Oxford University Press.

Mathers, C. D. (2002). Health expectancies: An overview and critical appraisal. In C. J. L. Murray, J. A. Salomon, C. D. Mathers, & A. D. Lopez (Eds.), *Summary measures of population health: Concepts, ethics, measurement and applications* (pp. 177–204). Geneva: World Health Organization.

Men, T., Brennan, P., Boffetta, P., & Zaridze, D. (2003). Russian mortality trends for 1991–2001: Analysis by cause and region. *British Medical Journal, 327*, 964–969.

Munnell, A. H., Hatch, R. E., & Lee, J. G. (2004). *Why is life expectancy so low in the United States?* (An Issue in Brief, No 21). Boston: Center for Retirement Research.

National Center for Education Statistics. (2007). *Comparative indicators of education in the United States and other G-8 countries: 2006*. Washington, DC: US Department of Education. http://nces.ed.gov/pubs2007/2007006_1.pdf. Accessed Oct 2009.

National Statistics. (2007). *Health expectancies at birth and at age 65 in the United Kingdom*. http://www.statistics.gov.uk/StatBase/Product.asp?vlnk=12964. Accessed Sept 2007.

Olshansky, S. J., Passaro, D. J., Hershow, R. C., Layden, J., et al. (2005). A potential decline in life expectancy in the United States in the 21st century. *The New England Journal of Medicine, 352*(11), 1138–1145.

Robine, J., Romieu, I., & Michel, J. P. (2003). Trends in health expectancies. In J. Robine, C. Jagger, C. D. Mathers, E. M. Crimmins, & R. M. Suzman (Eds.), *Determining health expectancies* (pp. 75–104). Chichester: Wiley.

Robine, J., Cheung, S. L. K., Horiuchi, S., & Thatcher, A. R. (2008). Is the compression of morbidity a universal phenomenon? In *Living to 100 and beyond*. Orlando, USA: Society of Actuaries. http://www.soa.org/library/monographs/retirement-systems/living-to-100-and-beyond/2008/january/mono-li08-04-cheung.pdf. Accessed Feb 2011.

Rowe, J. W., & Kahn, R. L. (1998). Successful aging (large print ed.). New York: Random House.

Sadana, R., Mathers, C. D., Lopez, A. D., Murray, C. J. L., & Iburg, K. M. (2002). Comparative analyses of more than 50 household surveys on health status. In C. J. L. Murray, J. A. Salomon,

C. D. Mathers, & A. D. Lopez (Eds.), *Summary measures of population health: Concepts, ethics, measurement and applications* (pp. 369–386). Geneva: World Health Organization.

Shaw, M., Dorling, D., & Smith, G. D. (2006). Poverty, social exclusion, and minorities. In M. Marmot & R. G. Wilkinson (Eds.), *Social determinants of health* (2nd ed., pp. 196–223). Oxford: Oxford University Press.

Shkolnikov, V., McKee, M., & Leon, D. A. (2001). Changes in life expectancy in Russia in the mid-1990s. *The Lancet, 357*, 917–921.

Spillman, B. C. (2004). Changes in elderly disability rates and the implications for health care utilization and cost. *The Milbank Quarterly, 82*(1), 157–194.

WHO (World Health Organisation). (2004a). *Life tables for 191 countries, world mortality in 2000*. http://www3.who.int/whosis/life_tables/life_tables.xls. Accessed May 2008.

WHO (World Health Organisation). (2004b). *World health report 2004*. http://www.who.int/whr/2004/en/. Accessed May 2008.

World Bank Group. (2004). *Classification of economies by income and region, 2000*. http://www.worldbank.org/depweb/english/beyond/global/classification.html. Accessed May 2008.

Part II
Communities, Families and Individuals

Chapter 6
Community Change

An increasing trend towards greater demographic and socio-economic diversity ... [has] brought about a more serious focus among scholars on how changing population patterns shape the vulnerability and resiliency of social systems

(Donner and Rodríguez 2008: 1089)

6.1 The Community Perspective

Population aging is well-recognized as a concern for national policy-making, yet its impact is at least equally important at regional and community levels. The national experience of aging is the combined effect of aging in every locality – where effects are more immediate and have conspicuous consequences for community life. Extremes in population aging, diverse trends and rapid change are all part of communities' experience of population aging, rather than the more regular accumulation of changes usually anticipated at the national level. In this chapter, the concept of community refers mainly to spatial units in which older people live and spend much of their time – for example, municipalities, suburbs, shires and counties. It also includes groups or networks of people with a shared interest or identity, especially ethnic communities whose members may be dispersed across a number of locations. Glimpses of the future of aging societies are available in communities that already have a substantial representation of older people. In 2008, 456 counties in the United States, mainly in rural areas, had 20% or more aged 65 and over, compared with 330 counties in 2000 (Rural Policy Research Institute 2011).

The community is not only the setting for most older people's lives but also the site for responses to demographic and social changes, through policies and programs. From the perspective of identifying implications of population aging at the community level, a priority is to understand current trends in the overall distribution of the aged population. This calls for information about the two main processes of change in the numbers of the aged in communities, namely 'aging in place' and net migration.

D.T. Rowland, *Population Aging: The Transformation of Societies*,
International Perspectives on Aging 3, DOI 10.1007/978-94-007-4050-1_6,

Aging in place has become a major focus for policy-making, while migration is the process responsible for many of the uncertainties and extremes in community change. The ensuing discussion of these two processes considers explanations of them from empirical studies and theories, such as the 'elderly mobility transition', together with their consequences, including the growth and aging of ethnic communities and greater population diversity within communities.

6.2 Aging in Place

Aging in place gives rise to much of the aging of local populations, because the majority of people stay in the same residence as they grow older. Older people generally prefer to remain in the same neighbourhood, if not the same residence, where they have lived for many years. Accordingly, the spatial distribution of people in their sixties in developed countries is likely to denote where the majority of the elderly will be living 20 or 30 years later. In regions of the United Kingdom, for example, observed and projected figures for 1975–2031 show aging in place predominating as the cause of changes in the numbers of the aged, except in the south of England where there are appreciable net migration gains of retirees (Rees 1992a: 222–223). Despite this broad pattern, in some regions migration before age 60 reduces or augments the potential for aging in place (ibid.: 223). As well, the details of migration flows of the elderly reveal complex shifts within and between regions (Raymer et al. 2007).

Because of people's attachments to particular localities and social networks within them, aging in place has become a policy principle internationally (see Chap. 12). This principle supports the goal of enabling older people to remain in their own homes, or at least in their home community, for as long as possible. National aging in place or 'community care' policies for the disabled aged emphasize the former, with movement from the home to a hostel or nursing home intended mainly a last resort. Aging in place policies can also extend to making the home community more livable, for instance through providing appropriate and affordable housing, accessible transport and support services, and opportunities for engagement in recreational activities, paid or volunteer work and life-long education. The World Health Organisation's (2007) publication *Global Age-Friendly Cities: A Guide* describes a fairly comprehensive set of features conducive to quality of life and positive experiences of aging in place (Box 6.1).

Evidence of the importance of aging in place is the low proportion of older people who change their place of residence during a fixed interval in a wide range of countries (Bell and Muhidin 2009: 34–40). For example, in the United States only 3.4% of people aged 65 and over lived in a different home in 2009 than they did in 2008 compared with 24% of people aged 20–34 (U.S. Census Bureau 2009a, b). At younger ages, major reasons for moving include marriage or partnering, tertiary education, employment and housing considerations, including the needs of a young family. All of these, apart from housing, have minimal relevance at older ages.

Box 6.1 Characteristics of Age-Friendly Cities

The World Health Organization's notion of age-friendly cities builds on its active aging framework (see Chap. 12). "Active ageing is the process of optimizing opportunities for health, participation and security in order to enhance quality of life as people age." In an age-friendly city, policies, services, settings and structures support and enable people to age actively by:

- recognizing their wide range of capacities and resources;
- anticipating and responding flexibly to aging-related needs and preferences;
- respecting their decisions and lifestyle choices;
- protecting those who are most vulnerable;
- promoting their inclusion in and contribution to all areas of community life.' (WHO 2007: 5).

Global Age-Friendly Cities: A Guide, is a valuable starting point for setting objectives in community planning for older people. It identified an extensive range of core characteristics of age-friendly cities, identified through consultations with 1,500 older people and 750 caregivers and service providers from 33 cities in 22 countries. The outline below indicates the scope of these characteristics (ibid.: 12ff.).

Age-friendly outdoor spaces and buildings: environment, green spaces and walkways, outdoor seating, pavements, roads, traffic, cycle paths, safety, services, buildings, public toilets.

Age-friendly transportation: affordability, reliability and frequency, travel destinations, age-friendly vehicles, specialized services, priority seating, transport drivers, safety and comfort, transport stops and stations, information, community transport, taxis, roads, driving competence, parking.

Age-friendly housing: affordability, essential services, design, modifications, maintenance, aging in place, community integration, housing options, living environment.

Age-friendly social participation: accessibility of events and activities, affordability, range of events and activities, facilities and settings, promotion and awareness of activities, addressing isolation, fostering community integration.

Age-friendly respect and social inclusion: respectful and inclusive services, public images of aging, intergenerational and family interactions, public education, community inclusion, economic inclusion.

Age-friendly civic participation and employment: volunteering options, employment options, training, accessibility, civic participation, valued contributions, entrepreneurship, pay.

(continued)

Box 6.1 (continued)

Age-friendly communication and information: information offer, oral communication, printed information, plain language, automated communication and equipment, computers and the Internet.

Age-friendly community and health services: service accessibility, offer of services, voluntary support, emergency planning and care.

At the same time, there are significant disadvantages of moving in later life, because of stresses and financial costs and breaking ties with the home and familiar environment. Financial losses and emotional strain can result if the shift proves unsatisfactory and a return to the place of origin is necessary. Moving is also an obstacle to social integration through the weakening of established interpersonal ties: cessation of accustomed participation in networks of relatives, friends and associates, and loss of engagement in the activities of recreational, cultural and religious organizations. Moreover modern transportation and the ability of many to afford travel costs enable older people to enjoy sojourns in a wide range of desirable destinations, without the upheaval of migrating. The trend in social policies towards providing care at home through domiciliary services further reduces the need for migration. Aging in place opposes unwanted residential shifts.

Despite the positive benefits of not moving, lack of choice – in terms of appropriate and affordable housing – or the absence of necessary assistance to relocate, create negative consequences of aging in place. Thus it can result in isolation and loneliness as well as deficits in support for independent living arising, for example, from the physical environment of the home and weaknesses in social networks (Sixsmith and Sixsmith 2008). Transport and isolation become issues of mounting importance for people who stay too long at the same address, such as when frailty and poor eyesight limit capacities to drive a car or use public transport. Lack of suitable housing choices within the neighbourhood constrains many of the aged who would like to move and helps to perpetuate under-occupancy of family homes by the older population as a whole. Tax concessions for older people purchasing a smaller home, or other incentives and assistance to downsize accommodation, can result in more efficient use of the housing stock. Yet, overall, not moving is generally much preferred because of the desire to continue to live in familiar surroundings, maintaining ties with family and friends.

The contribution of aging in place to the growth of the older population consists of the net outcome of gains in older age groups, from greater numbers growing older, and losses from deaths. These two processes cause considerable turnover in the membership of populations. Total numbers will vary little in a decade if deaths offset most of the new additions from people passing age 65, but continuing turnover inevitably alters the composition of the community's aged population. Turnover has impacts at several levels. For older individuals it can denote welcome additions to their ranks, for instance as new retirees join in activities and take up responsibilities, but it can also result in disagreeable or distressing alterations in the circle of relatives, friends and acquaintances. For organizations of the elderly it encompasses the continual gains and losses in membership that affect the maintenance of activities as well as people's sense of camaraderie and belonging. Additionally, for administrators and service providers, turnover augments difficulties in maintaining registers of clients and generates concerns about the viability of particular services and capacities to meet future demands.

6.3 Migration

At the same time, migration can heighten the pace and impact of population aging on communities, but it is invariably the most unpredictable component of community change. It calls for foresight and early responses to meet contingencies that may exceed local resources. Migration at all ages is often the main cause of demographic change and population aging at the community level. Migration causes population aging in communities, both through net losses of the young and through net gains of the aged, especially those under 75. Also, although most moves of older people are short-distance local migrations and reflect preferences to remain in familiar places, the less common process of internal migration in later life commonly targets selected retirement destinations where population aging may be greatly augmented and accelerated. Initially, population growth from retirement migration can transform and revitalize the socio-economic characteristics of the receiving population, especially if the arrivals are enterprising, active and financially secure. However, within 20 years a cohort of largely healthy, independent retirement migrants – who need housing and community infrastructure similar to that for other adult age groups – inevitably becomes a cohort with many disabled and widowed aged who have special housing and transport needs and substantial requirements for health and support services. In the long run retirement migrations can lead to returns to places of origin as well as other departures, for instance because of widowhood or insufficient medical and aged care services.

Besides local migration (within communities) and internal migration (between communities), international migration is also a significant factor in community change. International migration of labour is expected to have a continuing economic role in developed countries, as well as in ameliorating population decline and aging. Yet the consequences of international migration for communities and societies are enduring because former immigrants grow older and many have special needs in their later lives. In high immigration countries recognition of the need for culture-specific care has led to varied initiatives. These include ethno-specific nursing homes and other residential accommodation for particular groups, 'clustering' of small numbers from the same origins in mainstream institutions, domiciliary services for basic needs (e.g. meals and housekeeping) provided, where possible, by members of the same ethnic community, and other integrated 'packages' of necessary home-delivered care. In Australia the 'packages' may include home help, laundry, shopping and assistance with meals and bathing, which can serve as an alternative to placement in residential care (Rowland 2007).

Government support for home-delivered services accords with common ethnic community expectations of family care for their aged, and also facilitates widespread support for small or scattered groups. Nevertheless, cultural inappropriateness of services is often an obstacle to obtaining needed support, for instance because there are insufficient staff and volunteers who speak the community language. In the United States, such concerns have led to advocacy of 'cultural competence' among health workers and the drafting of national standards for culturally and linguistically appropriate services in health care. Cultural competence entails the ability of health care organizations and employees to understand and respond effectively to the cultural and linguistic needs of service recipients (Office of Minority Health 2000a, b). Inadequate funding and the complexity of responding to diverse and changing client populations are further impediments to achieving equity in access to needed forms of support.

Special needs among the ethnic minority aged originate from the importance to them of their distinctive cultural and social circumstances and associated incompatibilities with attitudes and expectations that influence the provision of mainstream care. Cultural differences in perceptions of successful aging, for example, are evident in that Americans associate it with self-sufficiency and living alone, while aged Chinese in Hong Kong associate it, not with self-sufficiency, but with their families' willingness to meet their needs. Moreover, to Americans self-perceptions of being "optimistic, courageous and motivated" signify successful aging, whereas to the Chinese aged it resides in others thinking of them as "tolerant and easy-going" (Torres 1999: 38). The way social groups perceive successful aging depends on their value system and it is misleading to interpret the perceptions of one culture according to the values of another. Whereas Chinese elderly may consider that being tolerant and easy-going are desirable characteristics in their own right, non-Chinese may interpret such characteristics mainly as practical adaptations to being coresident with younger relatives. Each interpretation arises from a different value orientation; the former conveys a preference for intergenerational living, the latter a preference for individualistic relations with others (ibid.: 38 and 46).

Nevertheless, the population at risk of having ethno-specific needs in aged care is likely to be a fraction of the total ethnic group, expect where immigration from a particular origin was short-lived and the whole group reaches advanced ages together. International migration of people in the young working ages over a limited interval ultimately generates some of the world's oldest ethnic group age structures. Instances include the age structures that evolved at destinations following the international migration of Eastern European refugees after the Second World War.

Planning provision for the aging of ethnic minorities calls for current and projected figures on their overall numbers, subdivided according to whether or not they require aged care, with the latter subdivided again according to whether or not they require culture-specific care. Membership of this last group depends on maintenance of the ethnic culture, the likelihood of which is especially high among those who have never become proficient in the national language of the country of settlement (Rowland 1999). Censuses and health surveys can provide the data needed for such estimates and projections, or they at least provide approximations that are far more informative than total population numbers.

Although much emphasis is given to disadvantages of being a member of an ethnic community, socially integrated former immigrants may enjoy the best of both worlds, benefiting from participation in their own ethnic group as well as in the structures and organizations of the society as a whole. There are also benefits for the aged generally accruing from information about the advantages for health and well-being bestowed by aspects of ethnic cultures, such as the Mediterranean diet. Clearly, policies for the aged in communities need to be made relevant and implemented for minorities as well as for the majority population. Some ethnic minority groups may have both a disadvantaged socio-economic position and a heightened risk of poor health, sometimes exacerbated by the additional effects of prejudice and racism (Shaw et al 2006: 210–212).

6.4 Explaining Migration

An initial theoretical framework for describing and explaining the migration of older people is Wilbur Zelinski's 'mobility transition'. In his pioneering paper on the hypothesis of the mobility transition Zelinsky (1971) proposed that, paralleling the demographic transition, there are patterned changes though time in rates of different types of movement, such as from the countryside to towns, between towns and between countries. Criticisms of the mobility transition have referred to its Eurocentric focus, overlooking the wide variations in the nature of mobility in countries around the world, as well as the lack of a close correspondence between modernization and certain types of mobility (Bell and Muhidin 2009: 52). Nevertheless, Zelinsky's ideas have stimulated ongoing interest in the characteristics of mobility patterns through time including those of the aged. Various researchers have contributed ideas concerning what Rogers (1989) originally termed the 'elderly mobility transition'. This has sought to provide a broad temporal perspective linking changes in migration with other trends in developed countries.

Currently, however, the existence of a common sequence of changes through time, relevant to different developed countries, is far from confirmed. The three stages of the elderly mobility transition (Box 6.2) derive from descriptions of different settings at different times and generalization has proven difficult because of "considerable inter-country variability" (Bean et al. 1994: 342). As in the demographic transition, diversity in national experience of the elderly mobility transition is to be expected because geographical, social, economic and cultural differences influence the causes of change. Longino and Warnes (2005: 539–541) accordingly restated the mobility transition in their research as "a stage model of types and destinations" with particular reference to retirement migration in English-speaking countries and France.

A development conducive to the supposed third stage (Box 6.2) is the emergence of vast urban regions, which are less dense and compact than single urban areas.

Box 6.2 The Elderly Mobility Transition

This idealized model for industrialized societies has three stages (Rogers 1989: 19; Bean et al. 1994: 341; Rees 1992b: 182; Longino and Warnes 2005):

1. Stage 1 (early twentieth century). Concentrations of the elderly emerge in rural regions of most intense outward migration of the young. Low levels of return migration occur at retirement, from urban areas to widespread rural areas of origin, reflecting rural ties.
2. Stage 2 The second stage brings a decline in return retirement migration to rural areas generally, because fewer urban dwellers have such ties, together with higher levels of retirement moves to selected, environmentally attractive destinations.
3. Stage 3 (late twentieth century). Emergence of more dispersed destinations in non-metropolitan areas as the attractiveness of the original retirement towns decreases, due to rising population numbers and property costs.

Some research suggested that, in the 1990s, the United Kingdom was in the third stage, the United States was approaching the third stage while Japan and Italy were probably still in the first stage (Bean et al 1994: 341). In Japan, there has been little retirement migration, but there have been rising rates of female migration after widowhood (Rees 1992b).

They consist of clusters of cities and towns linked together through commuting and other social and economic ties. Living within less densely settled parts of an urban region can bestow on residents the advantages of a large city (e.g. access to medical services, entertainment, and cultural and sporting activities), while lessening some its disadvantages (e.g. crowding, daily traffic congestion), and potentially expanding choices, including access to scenic and recreational environments. Living in an urban region also provides opportunities for day trips to enjoy the amenities of a range of different places without having to move. Thus short-term mobility within the same urban region enables some older people to achieve the goals of longer distance internal migrations without the upheaval that they entail. Seeking '*urbs in rure*' – places offering advantages of both urban and rural living – are major catalysts for migration in the early part of later life. The irony is that the more people pursue this quest, the faster the degradation of the environmental qualities that attracted them. Also there are doubts about whether opportunities for retirement migration will remain favourable. Obstacles include rising prices of housing and petrol, high rates of relationship breakdown and divorce, and a possible need to remain in the labour force for longer to finance a retirement income. In the United States, migration rates for all age groups have declined since about 1950, especially for moves over shorter distances (Wolf and Longino 2005:1). Nevertheless, rates of interstate movement of people aged 65 and over have remained unchanged, implying that long-distance retirement migration is continuing at previous levels (ibid.: 9).

Part III
Policies and Prospects

Chapter 12
Policy Concepts

History and experience tell us that major transitions in public policy usually occur at the confluence of a real or perceived crisis along with the availability of worked-out ideas and evidence, which find their time is come ... .

(Johnson 2005)

12.1 The Expanding Policy Agenda

In developed countries, current projections of unfavourable futures are the basis for forward-looking policy initiatives addressing a range of concerns including support for family formation, the provision of health services, care for the aged and the need to avert labour shortages and shortfalls in pension funding. In some developed countries there has already been a long period of development of strategies for adapting to population aging. Nevertheless, there are considerable variations in degrees of preparedness for change. Although policy making and program development entail detailed arrangements specific to national and local conditions, it is important that there be broader policy approaches and concepts to place details in perspective and define major goals. Policy concepts encapsulate a range of ideas and are important in establishing a foundation for the long-term courses of action so necessary to enabling timely responses and achieving equitable outcomes for successive generations. They include 'active aging', 'positive aging' and 'productive aging'. Their emergence reflects the challenges inherent in the new demography of aging and a deepening understanding of the consequences of aging for individuals and societies.

An important distinction is between policies to modify and ameliorate the course of population aging itself, and policies concerned with the welfare of older people. The former policies are underdeveloped and contentious because the most serious effects of population aging are only beginning to emerge, and because there is a widely-held belief that governments can do little to modify the determinants of

D.T. Rowland, *Population Aging: The Transformation of Societies*,
International Perspectives on Aging 3, DOI 10.1007/978-94-007-4050-1_12,

population aging. Welfare-related policies comprise an area of greater agreement and progress although policy choices involve compromises and uncertainties, for instance because of funding constraints and changes in societies through time. Policy-related concepts in the field of aging vary in meaning and currency and among policy makers there is no agreed set of concepts encompassing overarching goals for aging populations. Nevertheless there are a number of approaches which, together, go far towards focusing deliberations. This chapter is concerned with policy concepts developed through national and international debate and action.

12.2 Active Aging

The Second World Assembly on Ageing, convened in Madrid in 2002, was the venue for the launching of the World Health Organisation's (WHO) concept of 'active aging', which it had adopted in the late 1990s, replacing its previous emphasis on 'healthy aging.' The WHO defines active aging as "the process of optimizing opportunities for health, participation and security in order to enhance quality of life as people age." This was intended to be more inclusive than healthy aging, recognizing the range of factors affecting how individuals and populations age. At first sight 'active' might imply an emphasis on keeping physically active as a means of maintaining health and well-being, but its scope is much broader. Health, participation and security are the three pillars of a policy framework for active aging. The WHO adopted active aging to denote the process for achieving the vision that: "If ageing is to be a positive experience, longer life must be accompanied by continuous opportunities for health, participation and security" (WHO 2002: 12–13). Elaborating on this, the WHO's document on active aging presented at the Madrid conference stated:

> Active ageing applies to both individuals and population groups. It allows people to realize their potential for physical, social, and mental well being throughout the life course and to participate in society according to their needs, desires and capacities, while providing them with adequate protection, security and care when they require assistance.
>
> The word "active" refers to continuing participation in social, economic, cultural, spiritual and civic affairs, not just the ability to be physically active or to participate in the labour force. Older people who retire from work and those who are ill or live with disabilities can remain active contributors to their families, peers, communities and nations. Active ageing aims to extend healthy life expectancy and quality of life for all people as they age, including those who are frail, disabled and in need of care.
>
> "Health" refers to physical, mental and social well being as expressed in the WHO definition of health. Thus, in an active ageing framework, policies and programmes that promote mental health and social connections are as important as those that improve physical health status.
>
> Maintaining autonomy and independence as one grows older is a key goal for both individuals and policy makers. Moreover, ageing takes place within the context of others – friends, work associates, neighbours and family members. This is why interdependence as well as intergenerational solidarity (two-way giving and receiving between individuals as well as older and younger generations) are important tenets of active ageing. (ibid.: 12).

Major outcomes of active aging are expected to be: fewer premature deaths, fewer disabilities, greater prevalence of positive quality of life, higher participation in family, community, cultural, economic and political life, and lower costs related to medical treatment and care services (ibid.: 16). The breadth of the concept of active aging highlights the impossibility of finding a phrase that genuinely encapsulates the range of goals for policies to improve the experience of individual and population aging. Active aging is a convenient label for a set of goals rather than a phrase that immediately conveys a particular meaning or unified purpose. Although active aging seeks to be more inclusive than healthy aging, some researchers have used the latter with similar wide-ranging ramifications. Social participation, active engagement with life and maintaining independence for as long as possible are commonly recognized elements of healthy aging.

12.3 Positive Aging

'Positive aging' has also gained currency in describing national as well as community policy goals. The United Kingdom, for example, has had an 'Age Positive' strategy to promote the benefits of an age-diverse labour force with a focus on "potential, skills and ability, not age" (Department of Work and Pensions 2011). Under the United Kingdom's Employment Equality (Age) Regulations 2006 it is unlawful to discriminate against employees, job seekers and trainees on the grounds of age. Positive aging opposes ageism and the negative perceptions and behaviours that contribute to age-based social inequality, social exclusion and undervaluation of the present and potential contributions of older people. Like active aging it calls for recognition of older persons' rights as well as their needs, preferences and capacities. Implicit within positive aging is acknowledgment of the importance of the social and human capital inherent in older people and the notion their growing numbers constitute a resource for society. Overturning the stereotypes of the aged withdrawing from society and having a 'roleless role', it recognizes the contributions of 'retired' people together with the positive personal and societal benefits of continuing social engagement. As Minichello and Coulson (2005: 24) observed: "the binary of retiree versus worker is being challenged by the notion of 'active retirement' in which older people undertake various types of productive labour and retain active lifestyles."

Canberra, the capital of Australia situated in the Australian Capital Territory (A.C.T.), provides an example of a city where positive aging forms the basis of approaches to achieving quality of life for older people. The A.C.T. Government described positive aging as follows:

> Each person's ageing is influenced by many factors including socio-economic circumstances, life experience, gender, cultural background, education and general health. Positive ageing is about having a good quality of life through social relationships, a healthy lifestyle and feeling welcomed in the community. Positive ageing is about individuals, the community, businesses and governments valuing the experience of older people and providing them with opportunities to continue to contribute. It is also about seniors remaining involved, caring for themselves and each other and embracing this stage of their lives. (DHCS 2010: 4).

Canberra's positive aging plan has much in common both with the WHO's active aging concept and with the WHO's 'age friendly cities' framework (WHO 2007; see Box 6.1).

New Zealand's Positive Ageing Strategy, launched in 2001, promotes the participation and value of older people: it encourages them to be involved in their communities and recognizes the value of their knowledge, skills and experience. The strategy thus views the aged as a group with the right to be afforded dignity, and their expanding numbers as an asset. It is intended to provide a framework within which to understand and explore all policy implications for older people (Office for Senior Citizens 2001). Ten broad principles within the framework aim to guide the development of policies and services. Some resemble aspects of the WHO's active aging concept, notably health, participation and security, others reflect particular features of New Zealand society, especially its cultural diversity. For example:

- Empower older people to make choices that enable them to live a satisfying life and lead a healthy lifestyle;
- Provide opportunities for older people to participate in and contribute to family, whānau [extended family] and community;
- Reflect positive attitudes to older people;
- Affirm the values and strengthen the capabilities of older Maori and their whānau.

Cultural appropriateness, self-reliance, aging in place and productive retirement are integral to New Zealand's positive aging concept, further indicating that while the name designates a particular starting point or emphasis it also embraces a set of priorities common to other approaches:

> Positive ageing in general terms is a discourse that links to a number of influential discourses to do with healthy lifestyles, anti-discrimination and inclusion, productive and active lifelong citizenship, ethical work practices, fairness and social justice, and the need to disrupt some of the ideas that have constructed the ageing in negative or abject terms. (Minichello and Coulson 2005: 25).

12.4 Healthy Aging

Just as the WHO replaced healthy aging with active aging as a more encompassing term to denote the basis of a positive experience of later life, so too national policy makers now conceive of healthy aging as one component, albeit essential, of a broader policy agenda. For example, healthy aging figured prominently as one of several complementary policy options discussed in the Second Interim Report of Canada's Special Senate Committee on Aging (Carstairs 2008). Other options included elements of active and positive aging emphasizing active living and combating ageism, together with aging in place of choice – addressing housing and domiciliary support needs in light of preferences for remaining in the family home. Successive Australian governments have, similarly, included healthy aging in broad agendas for responding to population aging, although they have stated their agendas and the health policy priorities in different ways. Australia's healthy aging strategy was formerly intended to "optimise opportunities for people to have physical, social and mental well-being throughout their lives" (Andrews 2001: 36). It sought to improve health and well-being for all through early detection and treatment of disease and disability as well as through prevention. Despite promoting 'successful aging', the strategy was more universal in aiming to benefit the health of all groups, including those with disabilities and chronic illnesses – through progress and rehabilitation wherever possible. More recently, Australia's response to population aging has emphasized health as an economic priority to ensure workforce productivity, and healthy lifestyles and risk prevention as essential to economic and social participation, independence and quality of life. 'Preventive health' has been a particular focus to reduce the burden of disease due to obesity, tobacco and alcohol (Department of Health and Ageing 2008: 3–4).

Healthy aging initiatives include support for members of primary care professions – general practitioners and pharmacists – in early detection and management of conditions such as diabetes and asthma, and in counseling patients with key risk factors of smoking, poor diet, misuse of alcohol and lack of exercise. Another goal has been fostering social norms of active living, such as through the media and local communities (Andrews 2001: 40–41). Policies on healthy aging recognize that its foundation is in improvements among younger age groups, as Canada's Special Senate Committee on Aging commented: "The focus of healthy aging is not exclusively on seniors, but rather on the experiences through the life-course which result in a high quality of life in the senior years" (Carstairs 2008: 9). Healthy aging offers benefits for the quality of life of individuals and families as well as potential savings in health care costs through prolonging independent living. It can also enhance productivity through enabling more to remain longer in the labour force, or undertake voluntary work, or contribute to family and community support. The success of healthy aging depends partly on individuals taking greater responsibility for their own health throughout life, and partly on reinforcement through the actions of the health care professions, businesses, communities and governments (Andrews 2001: 36). For example, hazardous working and living environments, inadequate health services, scarcity of opportunities for social participation, and lack of information about healthy lifestyle choices and disease prevention – all have serious consequences for the health of individuals, but are largely beyond their control. A prosperous national economy is generally a prerequisite for addressing such issues in depth.

Despite its potential to extend healthy life expectancy and, for some, compress later life morbidity into a shorter period, the future extent of benefits accruing from the pursuit of healthy aging remains uncertain. A key question is: how prevalent can healthy lifestyles become? There have been optimistic estimates of the scope for future improvements such as findings that "perhaps as much as half of the functional decline associated with aging is the result of disuse and can be reversed by exercise" and "an estimated 80% of health problems associated with old age [can be] prevented or postponed primarily through lifestyle changes" (Andrews 2001: 42–43). Nevertheless the new 'epidemics' of obesity, diabetes, neurodegenerative diseases and depression demonstrate that reversals as well as advances are occurring. Similarly, evidence of morbidity reduction and a more informed public may conceal other disadvantages. Reliance on medications may substitute for lifestyle changes, survival rates from stroke may improve without any reduction in its incidence, while Barsky's paradox implies that unintended adverse consequences may flow from continually confronting people with information about diseases and health risks.

Even in high income countries a further persistent impediment to progress is social inequality. Socially and economically disadvantaged people are least likely to enjoy good health, are more likely to engage in behaviours that endanger their health and their lives, and may be least able to benefit from information about health risks. Understanding of health inequalities is a necessary basis for effective policies but should not delay action in targeting vulnerable groups whose needs are well-known – such as indigenous peoples and older immigrants. In a range of countries, mortality rates and disability rates are higher for older people in relatively disadvantaged circumstances (Howse 2005: 10).

Other obstacles to lifestyle reform abound: addictions, irrationality and hubris; stressful lives and difficulties in finding the time or motivation to maintain a healthy diet or take regular exercise; innate partiality for food high in fat, salt and sugar; the cost of fresh fruit and vegetables; advertising's endless exhortations to consume energy dense snacks and beverages; the ubiquity and immediate availability of tempting food in shopping malls; and the greater size – at a greater price – of what is promoted as 'a single serve', such as a 4 in. muffin, a 6 in. sausage roll, a 12 in. sandwich roll, a biscuit the size of a saucer, a bucket of popcorn, and a triple scoop of ice cream. These minimize chances for healthy lifestyles, especially among the young, the uninformed, the disadvantaged and other victims of contemporary advertising and lifestyle manipulation for commercial profit.

Thus some writers are doubtful that substantial lifestyle reform is possible, especially because many people find lifestyle modifications difficult to maintain. Other writers advocate more emphasis, not on lifestyle reform, but on the development of drugs and therapies to prevent or treat chronic illnesses that are largely unrelated to lifestyles – such as arthritis and Alzheimer's disease. Yet others consider that addressing social inequality in health should be a central objective of healthy aging, especially through reducing poverty, which underlies other types of disadvantage. It appears that the gap in death rates between upper and lower socio-economic groups has been widening – in the UK, Italy and the Nordic countries – because, although there has been a general improvement, the better off groups have benefitted more (Howse 2005: 7–10). Both the latitude for improvement in population health and the means of achieving progress will remain high priorities for research. Underscoring the urgency of this research is Johnson's (2005: xxiii) warning that "current health and social care systems will be undeliverable as the post-Second World War 'baby boomer' generations enter the Fourth Age." Better health for the aged could benefit society generally, even promoting social change towards making the concept of 'retirement' outmoded and fostering a redefinition of the Third Age: "…as we live in health for longer periods of time, it is conceivable that we will re-think the traditional school-work-retirement transitions to allow for greater flexibility at all stages of life" (Carstairs 2008: 8).

12.5 Human Capital and Productive Aging

Because retirement is an event, and sometimes a repeated event, rather than a life stage, it is important to consider the nature of the Third Age in relation to the use and further development of older people's human capital resources. Besides health, these resources consist of formal education and the knowledge, skills and experience that enable productive contributions to family and community life. As earlier chapters have discussed, contributing in these ways can be mutually beneficial for the society and for individuals – stimulating interests, expanding social networks and enhancing health, independence and self-esteem. Better nurturing of human capital resources are foreshadowed in the *International Plan of Action on Ageing* 2002 and in policy proposals from the OECD (see Box 16.1). They include extending working life and removing incentives for early retirement. Utilizing human capital and raising productivity are also priorities in seeking to reduce youth unemployment and involuntary early retirements, both of which further limit the average number of years spent actively in the labour force.

During the twentieth century, retirement became established as a life cycle event, as well as one commencing earlier. In the United States in 1890, 89% of males aged 55–64 were in the labour force, compared with 68% in 1990. Similarly at ages 65 and over the corresponding figures were 68% and 16% (Davis and van den Oever 1981). Early retirement, in conjunction with more years spent in formal education, resulted in many males spending less than half of their lives in full-time employment. Since 1990 there have been increases in labour force participation for both sexes at older ages in the United States, especially for women aged 55–64 (Fig. 12.1). At the same

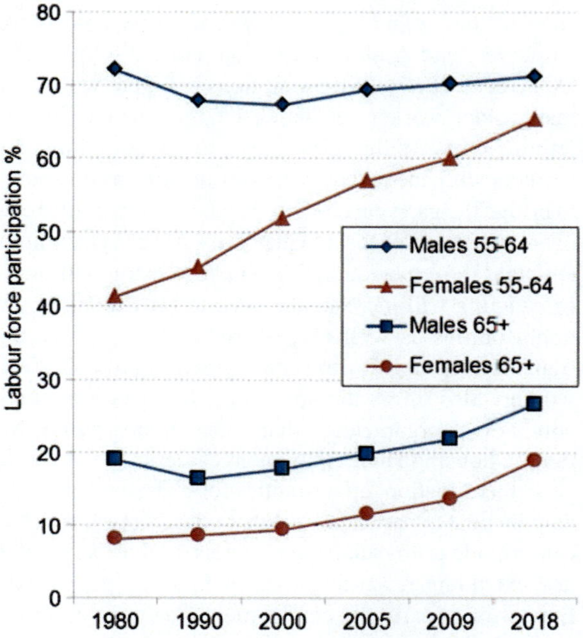

Fig. 12.1 Civilian labour force participation rates, United States 1980–2018 (Source: U.S. Census Bureau 2011)

time, however, overall male labour force participation has fallen, particularly at ages under 25: the participation rate for males fell from 76% in 1990 to 72% in 2009. In the female labour force, higher participation at ages at ages 25 and over has largely counterbalanced a fall in participation at younger ages (U.S. Census Bureau 2011). In some countries higher labour force participation of women has offset many of the male employment losses over time, but childbearing and family responsibilities, as well as high participation in tertiary education, commonly result in women having shorter working lives than males overall. Currently, governments are phasing in higher pensionable ages such as from 60 to 65 in Japan and Germany, and from 65 to 67 in the United States (2000–2027) (United Nations 2004: 50). Also occurring is an equalizing of the retirement ages for men and women (United Kingdom), tightening of eligibility requirements for full benefits (France) and reductions in full benefit levels (France, Germany, Italy, Japan, United Kingdom) (ibid.: 50).

A focus on human capital characterizes another recent policy-related concept, namely 'productive aging', which is concerned with the benefits for society and the individual of work of economic or social value. Although it has no widely accepted definition, discussions of productive aging place emphasis on the aged as contributors to society rather than merely consumers and dependants. This envisages contributions of economic value, whether paid or unpaid, and engagement in activities of wider benefit than just for the individual's own well-being (Bass and Caro 2001; Kaye et al. 2003). Productive aging thus has affinities with the social capital perspective in which civic engagement is a key feature.

The productivity of workers is far less a function of age than of skills and health status, although experience and lower rates of absenteeism and turnover can reinforce productivity in an older labour force. A key distinction is between the aged with motivation and high functional capacity and those without. A Dutch study showed that cognitive and physically based skills were valued more highly – by employers and employees, young and old – than reliability and commitment (Van Dalen et al. 2010). Past promotions to higher wage and salary levels, however, make many older workers a manifest target for employers wishing to cut staffing and training costs. Without incentives to stay, or with positive incentives to leave, many workers with adequate savings retire early, as do some others with sufficient savings to bridge the gap years before they become eligible for an age pension. Unfortunately over-estimation of the adequacy of savings is a common problem among early retirees, as is the expectation that an age pension will fund more than essentials. Labour force aging will increase the need to provide older workers with access to training opportunities as well as with more flexible working hours to facilitate gradual transitions from full-time employment (Andrews 2001: 22). Training of mature age workers also offers the advantage that they are more likely to stay with the same company or organization than more mobile young workers at early stages of their family, housing and employment careers.

In this situation, opportunities for phased retirement, through a shift from full-time employment to part-time, together with labour force re-entry to part-time employment, can provide a financial safety net as well as ways of increasing labour productivity and extending rewarding uses of time for people who wish to continue working. Expansion of part-time employment is a means of implementing recommended policy initiatives of raising labour force participation rates, removing disincentives for working beyond retirement age, and achieving flexible retirement arrangements (ibid.: 12–13). Many people would prefer to engage in part time work after they retire (see Uhlenberg 2000: 265) but there are major disincentives to taking up further paid employment, even for those most interested. Especially important in some countries is means testing of age pensions, despite the frugal life styles that they often entail. People who earn supplementary income may risk losing part or all of their age pension. For example, in Australia from March 2011 single age pensioners could earn no more than $146 per fortnight before the maximum fortnightly age pension of $670.90 was reduced by 50c for every dollar earned, even though the pension was no more than a poverty-prevention payment (see Centrelink 2011a). Loss of opportunities to increase personal income not only places a low ceiling on the material living standards of pensioners but also obstructs their labour force productivity and the benefits that their greater purchasing power and expenditure might bring. In Australia, an incentive for longer labour force participation is the 'Pension Bonus Scheme' which pays a modest tax-free lump sum to people who stay in the labour force for a minimum of 1 and a maximum of 5 years beyond the pensionable age, provided they register for the bonus in advance and are eligible for an age pension when they retire.

Income tests for pensions also leave many older people 'assets rich but cash poor', because home owners often have little or no private sources of income. Drawing down the wealth invested in the family home, such as through a reverse mortgage may also affect pension eligibility while having no advantage in terms of supplementing labour force productivity. Although publicly funded pensions should target the needy and should not supplement substantial private incomes, the rising representation of the aged in national populations calls for reform of pension eligibility rules. Incentives to engage in productive employment, and flexibility to allow pensioners access to more than a subsistence level income, would benefit older individuals, their families and communities.

For people with an adequate retirement income and no desire to continue paid employment, volunteering offers advantages for productive activity, as well as for social capital formation (see Chap. 7). Some suggested ways of raising participation in volunteering are through tax incentives, arrangements to match individuals appropriately with the needs of organizations, and adequate funding of volunteer organizations – to maintain continuity in their work as well as to ensure a sound infrastructure for recruitment and training. However, engaging in voluntary work is often a continuation of patterns established earlier in life, rather than a new type of behaviour arising conspicuously after retirement (Carstairs 2008: 11–12).

12.6 Aging in Place

The final policy concept here is aging in place, referred to also in Chap. 6. This is relevant to the whole of later life but it gains added importance when older people have less independence, or more constraints on their decision-making. The *International Plan of Action* 2002 made aging in place one of its objectives, declaring that “Ageing in one’s community is an ideal in all countries” (United Nations 2002: 34–5). Aging in place aims to enable people to remain in familiar surroundings as they age, rather than having to adjust to new and unfamiliar environments when they confront poorer health or widowhood. It can also extend to institutional and supported residential settings when attention is given to enabling residents to stay longer in their present quarters, rather than being required to move to less independent or higher care accommodation upon suffering a temporary setback or reaching age 80 or another specified birthday. Government funding arrangements based on disability levels can facilitate this (Gibson et al. 2002). Aging in place is a significant advance beyond the practice of prematurely relocating older people in hostels and nursing homes. In some countries levels of institutionalization of older people have declined as a result of this approach.

Aged care policies complementing aging in place are known alternatively as ‘community care’ or ‘home and community care’ policies, although ‘community care’ sometimes denotes primary health care policies. The types of domiciliary support provided can include basic services such as meals, housekeeping, transport and home maintenance for people requiring only moderate levels of assistance. Other clients, young or old, with higher care needs may receive a range of additional nursing, health care and respite care services. Provision of the higher levels of support enables disabled people and their carers to be more independent, to remain in their own homes and to avoid or delay admission to a nursing home. As well as benefitting individuals, home and community care provides flexibility in meeting needs and is a means of reducing the overall public cost of care. Institutional care is most expensive because of the finances required for building and maintaining a residential structure, as well as providing round-the-clock staffing every day of the year.

The potential disadvantage of home and community care is the imposition of continuing and under-recognized demands on the family members who act as ‘gatekeepers’ for home-delivered services and provide the supplementary 24 hour backup so necessary in many cases. This burden is most acute where expectations that the family, especially female relatives, should perform the ‘traditional’ function of supporting frail or disabled aged relatives, whether within cultural groups where such views are dominant, or in entire societies where community services are scarce or unaffordable.

Respite care is an important adjunct to community services, enabling carers to have some time for themselves, but the number of hours of respite care available is usually low. The Italian government has sought to provide ‘day hospital’ care in conjunction with increasing an emphasis on home care. Cash payments to compensate carers for their work and other earnings foregone tend to be modest or non-existent. Australia, Canada and Sweden have provided such allowances (United Nations 2004: 55–56). In Australia from March 2011 the Carer Allowance was A$110 per fortnight (Centrelink 2011b). Most countries have shown considerable reluctance to pay cash benefits to carers through the social security system. Instead there has been a rising tendency on the part of governments to emphasize the role of the family in providing care as an alternative to formal and institutional care (United Nations 2004: 55–6). This is regressive policy which can lead to long-term responsibilities and burdens which in the past only a minority of families faced – because relatively few people survived to advanced ages. Also, childlessness among the elderly together with sole parenthood, remarriages and other serial relationships among the younger generation combine to make family care far from universally accessible or preferred. In enabling aging in place, the family is a significant complement to community care rather than an alternative.

Another small-scale source of community support for the aged and their carers is Time Bank membership which allows people to earn credit, in terms of hours worked in providing support to others, which can be redeemed later through receiving an equivalent number of hours of support from members of the Time Bank. For example, people might earn credit in their Third Age and redeem it in their Fourth Age, or in a relative's Fourth Age. Time Banks have been established in 22 countries including the USA the UK, China and Japan. In Japan they serve important functions additional to exchanging support, notably through facilitating new friendships and giving people a sense of purpose and fulfilment (Miller 2008).

12.7 Conclusion

Aging in place and the other concepts provide bases for contemporary social policies concerning aging individuals and societies. All are relevant to reducing the public and private costs of aging. Contemporary policy concepts vary in their emphases but together they demonstrate recognition of the need for substantial policy agendas. These agendas extend well beyond those of earlier approaches in which dependency and custodial care were key concerns. Together with the 2002 *International Plan of Action*, discussed in Chap. 13, concepts such as active aging go far towards defining the scope of needed reforms, although they only partially address economic aspects. It is significant that neither the Plan of Action nor the main policy concepts advocate family care for the aged as the fundamental response to rising levels of aging. This contrasts with policies that some governments in Europe are promoting, but which fail to recognize that the prospective challenges for family care have never been so great, given the greater numbers of the aged surviving longer and the vulnerability of the family. Humane outcomes for the aged and their carers will depend more than ever on a significant sharing of responsibilities between families, communities, organizations and the state.

References

Andrews, K. (2001). *National strategy for an ageing Australia: An older Australia, challenges and opportunities for all.* The Hon Kevin Andrews MP, Minister for Ageing. Canberra: Commonwealth of Australia.

Bass, S. A., & Caro, F. G. (2001). Productive aging: A conceptual framework. In N. Morrow-Howell, J. Hinterlong, & M. Sherraden (Eds.), *Productive aging: Concepts and challenges* (pp. 37–78). Baltimore/London: The John Hopkins University Press.

Carstairs, S. (2008). *Issues and options for an aging population*, Special Senate Committee on Aging, Second Interim Report. Ottawa: The Senate.

Centrelink. (2011a). *Payment rates*. http://www.centrelink.gov.au/internet/internet.nsf/payments/age_rates.htm. See also: Income test for pensions. http://www.centrelink.gov.au/internet/internet.nsf/payments/chartc.htm. Accessed Apr 2011.

Centrelink. (2011b). *Carer allowance (caring for a person 16 years or over)*. http://www.centrelink.gov.au/internet/internet.nsf/payments/ca_adult_rates.htm. Accessed Apr 2011.

Davis, K., & van den Oever, P. (1981). Age relations and public policy in industrial societies. *Population and Development Review, 7*, 1–18.

Department of Health and Ageing. (2008). *Ageing and aged care in Australia*. Canberra: Commonwealth of Australia.

Department of Work and Pensions. (2011). *Age positive*. http://www.dwp.gov.uk/age-positive/. Accessed Apr 2011.

DHCS (A.C.T. Department of Disability, Housing and Community Services). (2010). *ACT strategic plan for positive ageing 2010–2014: Towards an age-friendly city*. Canberra: DHCS. http://www.dhcs.act.gov.au/__data/assets/pdf_file/0003/104889/Strategic_Plan_for_Positive_Ageing.pdf. Accessed Aug 2010.

Gibson, D., Rowland, F., Braun, P., & Angus, P. (2002). *Ageing in place: Before and after the 1997 aged care reforms* (AIHW Cat. No. AUS 26). Canberra: Australian Institute of Health and Welfare.

Howse, K. (2005). Policies for healthy ageing. *Ageing Horizons*, Issue 2, 3–15.

Johnson, M. L. (2005). General editor's preface. In M. L. Johnson (Ed.), *The Cambridge handbook of age and ageing* (pp. xxi–xxvi). Cambridge: Cambridge University Press.

Kaye, L. W., Butler, S. S., & Webster, N. M. (2003). Toward a productive ageing paradigm for geriatric practice. *Ageing International, 28*(2), 200–213.

Miller, E. J. (2008). *Both borrowers and lenders: Time banks and the aged in Japan*. Ph.D. thesis in Sociology, College of Arts and Social Sciences, Australian National University, Canberra.

Minichello, V., & Coulson, I. (Eds.). (2005). *Contemporary issues in gerontology: Promoting positive ageing*. Sydney: Allen and Unwin.

Office for Senior Citizens. (2001). *New Zealand positive ageing strategy*. Wellington: Office for Senior Citizens, Ministry of Social Development. http://www.msd.govt.nz/about-msd-and-our-work/publications-resources/planning-strategy/positive-ageing/index.html. Accessed June 2011.

Uhlenberg, P. (2000). Introduction: Why study age integration? *The Gerontologist, 40*(3), 261–266.

United Nations. (2002). *Report of the second world assembly on ageing, Madrid, 8–12 April 2002* (A/CONF.197/9). New York: United Nations.

United Nations. (2004). *Policy responses to population decline and ageing* (Population bulletin of the United Nations, Special issue nos. 44/45). New York: United Nations.

U.S. Census Bureau. (2011). *The 2011 statistical abstract*. Civilian labor force and participation rates with projections. http://www.census.gov/compendia/statab/cats/labor_force_employment_earnings/labor_force_status.html. Accessed June 2011.

Van Dalen, H. P., Henkens, K., & Schippers, J. (2010). Productivity of older workers: Perceptions of employers and employees. *Population and Development Review, 36*(2), 309–330.

WHO (World Health Organisation). (2002). *Active ageing: A policy framework*. Geneva: WHO.

WHO (World Health Organization). (2007). *Global age-friendly cities: A guide*. Geneva: WHO.

Chapter 13
Policy Responses

> *... inadequate national revenue generation and collection, combined with new challenges ... arising from demographic changes and other factors, jeopardize the financing of social services and social protection systems in many countries.*
>
> (United Nations 2002: 32)

13.1 The Second World Assembly on Ageing

The body of advice emerging from international forums and the work of international organizations is a vital setting for discussion of national policy responses to population aging, given that it is informed by research, substantial experience and varied perspectives. International effort has led to broad policy approaches with wide relevance. A major statement of these is the *Political Declaration of the Second World Assembly on Ageing*, formulated in Madrid in 2002 under the auspices of the United Nations (2002: 1–4). This built on work initiated 20 years earlier in Vienna at the First World Assembly on Ageing, which focused world attention on aging as a major phenomenon and drafted the first *International Plan of Action on Ageing*.

The Political Declaration set out guiding principles in 19 paragraph-length Articles. Several Articles (1, 3 and 19), of the Declaration emphasized a commitment to the development of 'a society for all ages' – which had been the theme for the 1999 International Year of Older Persons (ibid.: 7). The Declaration also called for "actions at all levels, including national and international levels, on three priority directions: (i) older persons and development, (ii) advancing health and well-being into old age, and (iii) ensuring enabling and supportive environments" (ibid.: 1):

> The priority directions are designed to guide policy formulation and implementation towards the specific goal of successful adjustment to an ageing world, in which success is measured in terms of social development, the improvement for older persons in quality of life and in the sustainability of the various systems, formal and informal, that underpin the quality of well-being throughout the life course. (ibid.: 8).

D.T. Rowland, *Population Aging: The Transformation of Societies*,
International Perspectives on Aging 3, DOI 10.1007/978-94-007-4050-1_13,

The Articles of the Political Declaration elaborated on these priorities. The content of many of the Articles overlapped with that of several others and covered a broad agenda, among which the most prominent are outlined below. While headings are used here for summary purposes, they were absent from the Political Declaration. Also, the Declaration evidently avoided ambiguity and controversy about the meaning of summary concepts by making no specific reference to any, apart from 'quality of life'.

1. *Social Integration*. Eliminate age discrimination and neglect, promote increased opportunities for older persons to remain independent, realize their potential and participate fully in all aspects of life (Articles 2, 5, 6, 12 and 14). While the Declaration did not mention 'social integration', the ideas here relate to this important concept in social gerontology.
2. *Health*. Seek to achieve realization of the right to enjoyment of the highest standard of physical and mental health. This requires action in other social and economic sectors besides the health sector, as well as policies concerned with care and treatment, supportive environments and promotion of healthy lifestyles (Articles 5, 6 and 14).
3. *Productivity*. The Declaration emphasized the great potential contribution of older people to future development. It stated that they should have the opportunity to work for as long as they wish and are able to, in satisfying and productive work, continuing to have access to education and training programmes. The Declaration also recognized their important role as caregivers (Articles 10, 12 and 14).
4. *Quality of life*. "Concerted action is required to transform the opportunities and the quality of life of men and women as they age and to ensure the sustainability of their support systems." A gender perspective is needed in all policies and programmes as well as particular concern for the vulnerable aged generally and for those in insecure or life-threatening situations (Articles 5, 6, 8, 9 and 12).
5. *Social and economic development*. Aging needs to be included in development agendas, in strategies to eradicate poverty and in efforts to achieve full participation in the global economy of all developing countries (Articles 5, 7, 8 and 10). Vital in developing countries is concern for social and economic development to alleviate hardship among the aged and the population generally.
6. *Implementation*. Governments bear primary responsibility for leadership on aging matters and for ensuring access to basic social services. Progress in the priority directions requires research to assist in the formulation of policies and calls for intergenerational solidarity, together with support for families, volunteers, communities, organizations, corporations, workers, educational and religious institutions and the media. The roles of the United Nations and international co-operation are also vital (Articles 4, 6, 11, 13, 15, 16, 17 and 18). The range in the types of support regarded as necessary to the implementation of policies highlights the importance of action at different levels within societies.

Although the above summary is not exhaustive, it illustrates that there were major common concerns ranging across the 19 Articles. The policy principles endorsed at the Second World Assembly on Ageing reflect many ideas from academic research and national policies. Important features of the Political Declaration are the emphasis on quality of life and social and economic development, placing the needs of aged women, as well as men, in developing countries prominently on the international agenda. Policies to address the welfare needs of the aged in developing countries confront by far the greatest obstacles because of widespread poverty and the impact of migration and urbanization, which have separated relatives and weakened traditional support systems.

13.2 The Madrid International Plan of Action

The Political Declaration's Articles and three priority directions provided the framework for the Madrid conference's second major policy document: *The Madrid International Plan of Action, 2002* (United Nations 2002: 5ff.). This set out recommendations for action together with strategies for implementation and follow-up. Summary concepts were again largely absent from this document, apart from quality of life and aging in place (see ibid.: 32 and 34). The document lists 239 recommendations for action to guide policy making and program development. These were arranged under the headings of the three priority directions stated in the Political Declaration (Box 13.1). The Plan of Action was intended to be "a practical tool to assist policy makers to focus on the key priorities associated with the consequences of individual and population aging" (ibid.: 7). It also provides direction and standards for groups and organizations seeking to promote the welfare of older people and timely adjustment to demographic and social changes. A substantial number of the suggested actions are especially relevant to developing countries.

Box 13.1 Overview of the Madrid International Plan of Action on Ageing, 2002

Older Persons and Development

1. *Active participation in society and development.* Enable older persons to participate actively in society and continue contributing to it, e.g. through preventing discrimination and recognizing roles in care of family, productive subsistence work and voluntary activities and enabling their participation in decision making. Create an enabling environment for volunteering at all ages and promote civic and cultural empowerment.
2. *Work and the ageing labour force.* Enable older persons to continue income generating work as long as they want and can do so productively. Reduce incentives for early retirement and disincentives for working longer.

(continued)

Box 13.1 (continued)

3. *Rural development, migration and urbanization.* Ageing is marked in rural areas of developing countries due to an exodus of young adults, leaving older persons without traditional family support. The urban setting is less conducive to sustaining traditional family support networks. Rural and urban aged need improved living conditions and support.
4. *Access to knowledge, education and training*: to ensure productivity and employment of older persons and utilization of their potential and expertise.
5. *Intergenerational solidarity*, e.g. promote understanding of ageing through public education, and mutual support between generations as a key element for social development.
6. *Eradication of poverty, with particular attention to the needs of women and the disabled, e.g.* through enhanced international cooperation.
7. *Income security, social protection/social security and poverty prevention*, e.g. place pension systems on a sound financial footing, implement policies ensuring adequate economic and social protection during old age.
8. *Emergency situations*, e.g. equal access of older persons to care during and after natural disasters and other humanitarian emergencies.

Advancing Health and Well-Being into Old Age

1. *Health promotion and well-being throughout life*, e.g. through poverty eradication, addressing environmental and other risk factors, taking action to reduce smoking and alcohol abuse, promoting healthy lifestyles and social and civic participation of older people, achieving access to clean water, adequate nutrition and affordable dental services.
2. *Universal and equal access to health care services*, e.g. promoting access for people who are poor or in rural or remote areas, strengthening primary health care and long-term care services.
3. *Older persons and HIV/AIDS.* Improve support for older people with HIV/AIDS or who are caregivers for infected or surviving family members.
4. *Training of care providers and health professionals*, e.g. expand educational training in geriatrics and gerontology and provide informal caregivers with access to information and basic training.
5. *Mental health needs of older persons,* e.g. develop national and local strategies to improve prevention, treatment and training, as well as long-term care and respite care.
6. *Older persons and disabilities.* Develop national and local policies and programmes for the treatment and prevention of disabilities, e.g. housing options that facilitate independence, encourage provision of rehabilitation and integration into society of older persons with disabilities.

(continued)

Box 13.1 (continued)

Ensuring Enabling and Supportive Environments

1. *Housing and the living environment* – important to health and well-being. Promote ageing in place e.g. through initiatives in relation to integration of the aged and suitable, affordable housing; improve access to affordable transportation.
2. *Care and support for caregivers* – most care is informal, even in countries with well-developed formal care policies. Policy objectives of community care and ageing in place can overburden family caregivers if community care is under-resourced. A continuum of affordable care options is desirable, together with support for the care giving role of older persons, particularly older women, e.g. through respite services.
3. *Neglect, abuse and violence.* Work to prevent abuse, consumer fraud and crimes against older persons. Women are especially disadvantaged because of lack of access to economic resources, lack of education and minimal participation in decision making. Take action through education and legislation.
4. *Images of ageing.* Promote a positive view of ageing, opposing the portrayal of older persons as a drain on the economy with escalating needs for health and support services, e.g. through enhancement of public recognition of the contributions of older persons.

Source: United Nations (2002: 9ff).

Kofi Annan, the then Secretary-General of the United Nations, summarized the overriding objectives of the Plan of Action as follows:

> We need to recognize that, as more people are better educated, live longer and stay healthy longer, older persons can and do make greater contributions to society than ever before. By promoting their active participation in society and development, we can ensure that their invaluable gifts and experience are put to good use. Older persons who can work and want to should have the opportunity to do so; and all people should have the opportunity to continue learning throughout life.
>
> By creating support networks and enabling environments, we can engage the wider community in strengthening solidarity between generations and in combating abuse, violence, disrespect and discrimination against older people.
>
> By providing adequate and affordable health care, including preventive health measures, we can help older people maintain their independence for as long as possible. (United Nations 2002: 67).

The wide scope of the Plan of Action ensures its general relevance to both developed and developing countries. Yet the emphasis governments and organizations give to particular aspects will inevitably vary between countries because of different national situations related, for instance, to economic development, historical traditions, social diversity and the level of population aging. Thus some concerns raised in the Plan of Action – such as rural development, disaster relief, HIV/AIDS and access to clean water – are scarcely mentioned in policy agenda's for the aged in developed countries where priorities are usually less dire and basic. This raises the question of whether it is possible to do justice to identifying all priorities and major policy objectives in a single global agenda which must give due regard to national diversity and political differences on particular issues.

In the deliberations of the Madrid Assembly there was little emphasis on the need for early interventions and proactive planning to prevent difficulties from cumulating. Both are important, for example, in ensuring the sustainability of health care systems and income maintenance for the aged, as well as in addressing demographic concerns. As noted earlier, there was also minimal reference to the concepts, reviewed in Chap. 12, that are prominent in other international policy agendas and in the goals of national governments – even though they relate closely to different aspects of the Political Declaration and the Plan of Action. Similarly, approaches to economic policies for aging societies were not discussed in any detail. Instead there was a more general emphasis on the need for economic development and eradication of poverty in developing countries. The OECD has given greater attention to economic policies, especially those most relevant to its member countries.

At the Madrid Assembly, the NGOs' main criticism of the Plan of Action referred to a lack of "a strong follow-up and effective monitoring process" (Kirakosyan 2002). Nevertheless, the Plan of Action included a section on this (United Nations 2002: 39–42) and the Social Integration Branch of the United Nations Department of Economic and Social Affairs has assumed responsibility for overseeing implementation (Social Integration Branch 2011). Since the Assembly, United Nations organizations have convened a number of regional conferences focussing on the implementation of the Plan of Action (see Chap. 15). Also, in 2008, the U.N.'s Department of Economic and Social Affairs (UNDESA 2008) published a *Guide to the National Implementation of the Madrid International Plan of Action on Ageing* (see Chap. 15).

13.3 Population Policy

A conspicuous absence from the deliberations at Madrid was any attempt to address the question of modifying the course of national demographic trends. Even with conscientious efforts to implement the Plan of Action, in many countries the long-term sustainability of policies depends on restraining rapid population growth, or rapid population decline, and keeping the level of aging within bounds. By their nature the demographic forces involved in population aging and population growth are long-term and cumulative. The means of responding to them need to be proactive and sustained. Moreover, the characteristics of the new demography have heightened the consequences of demographic change beyond previous expectations, necessitating far-reaching adjustments. Internationally, the most serious consequences of population aging will not become conspicuous until the 2020s, but by then it will be too late to initiate gradual adaptations. Some major initiatives, including restructuring of pension and health care systems, must anticipate future developments well in advance to enable an undisruptive changeover to new arrangements:

> Over the coming decades, the decisive shift to an older age structure in Europe will challenge social security and health systems, may hinder productivity gains, and could affect global competitiveness and economic growth. It could also strain relations among generations, particularly between those who are on the contributing and receiving ends of public transfer programs. It may also diminish social cohesion, particularly if increasing labor demand leads to substantial immigration from other cultures. (Lutz et al. 2004: 306).

Foresight is possible in policy making for aging societies to the extent that demographic changes are more predictable for older groups than for younger ones, and population projections and statistical models permit the analysis of future scenarios and potential policy options. Exploration of policy alternatives is important because strategies that are efficient and beneficial in one context may become wasteful and burdensome in another. The best-known example is that the viability of the earliest pension schemes depended substantially on the fact that only a small proportion of people lived to the pensionable age and even fewer lived long thereafter. Today, rising life expectancies and longer life in receipt of a pension are compelling more governments to consider means testing of pensions, higher ages of eligibility and alternatives to the taxation system as the funding source for retirement incomes. In the current period of ongoing change, policy reform needs to be a continuing process.

Although the Madrid Political Declaration referred to the need for sustainable social support, the statement did not mention 'sustainable development', nor did it refer to demographic aspects of sustainability. These have been subjects of other forums, including the World Population Conferences held in 1974 (Bucharest), 1984 (Mexico) and 1994 (Cairo), but demographic objectives have remained controversial in international policy debate. The nature of equitable and sustainable rates of fertility and population growth, so important in restraining expansion in the percentages and numbers of older people, remain matters of judgment for each society rather than open to international consensus.

Furthermore, population policy has reached an impasse where: (i) it is seen largely as an instrument for addressing rapid population growth in developing countries, rather than demographic concerns more generally; (ii) the effectiveness of family planning and reproductive health programs in some countries has fuelled a belief that further progress is assured, whereas many countries face decades of rapid population growth unless there is sustained international support for social and economic development and policy implementation; (iii) some associate population policies with repressive regimes although, in many cases, the policies have been a means of reducing mortality and improving living conditions, especially for women and children; (iv) the emergence of below replacement fertility has also created an impression that such policies can have unintended consequences, whereas very low fertility is an outcome of social and economic changes.

Writers offer varying definitions of population policy, but a common meaning is: "deliberately constructed or modified institutional arrangements and/or specific programs through which governments influence, directly or indirectly, demographic change" (Demeny 2003). The aims of a population policy are defined in demographic terms. A population policy entails active or proactive interventions into demographic trends, where necessary, to avoid detrimental outcomes. Thus seeking to influence the shape of the age profile through policies supporting higher or lower birth rates, and maintaining labour force numbers through immigration, are elements of a population policy irrespective of whether governments describe them as such.

Some of the strongest advocates of national population policies in developed countries are environmentalists who regard curbing population numbers as necessary to address climate change, resource depletion and environmental deterioration. Yet substantial population decline and high levels of population aging are likely to have adverse consequences without lessening other threats to ecological sustainability, including inadequate environmental protection legislation and high per capita resource use. Economies that become burdened with severe effects of labour force decline and population aging will have fewer resources to devote to matters other than immediate needs. When governments have sought to take effective action they have preferred to focus on the causes of particular environmental problems or on fostering technological changes, rather than pursuing an untargeted and uncertain demographic 'solution'. The fall in the population growth rates of most developed countries since the mid-1960s has done nothing to ease concerns about the environment as evidence of land degradation, extinctions, and global warming has continued to accumulate.

While many developing countries have adopted national population goals focusing on specific aspects of their populations, such as high birth rates, developed countries have had broader demographic concerns that are relevant to a number of areas of public policy. These have been addressed mostly through incorporating demographic considerations into social and economic policies, rather than through specific demographic policies. Demographic goals tend to be multi-purpose and not reducible to straightforward targets. Also, responsibility for population-related initiatives in developed countries is mostly distributed through various portfolio areas as well as through different levels of government, from local to national.

13.4 Interventions

Progress towards mitigating adverse consequences of population aging will depend upon: "an exploration of the demographic choices open to societies that aim not just to muddle through" (Demeny 1988: 241). Greater attention to this is right for the time because of the advantages of overseeing a range of demographic changes, and the negative consequences of not doing so. Policy interventions in developing countries have supported falls in total fertility rates from 6.0 to 2.0. Now initiatives are needed to raise total fertility rates in some developed countries by between 0.2 and 0.6 – a seemingly more difficult undertaking. Persistence of these relatively small differences is capable of transforming prospects for societies. The ethical focus for fertility-related policies is to enable women to achieve their family building goals as social and economic changes alter their circumstances. Having more than the wanted number of children has been a policy concern in developing countries. An emerging new focus in developed countries is the phenomenon of having fewer than the desired number of children.

By 2050, the United Nations (2004: 6) projects that 58 countries may be experiencing population decline, compared with 21 at the start of the century. United Nations' surveys have shown that an increasing number of governments consider that their population growth rate is too low and recognize aging as a major concern. Yet there has not been commensurate action, partly because a fall in the size of the total population and the labour force has not been perceived as a problem over the short term in countries with high levels of unemployment (ibid.: 12 and 24). In some countries non-intervention has been the stated position, even where the rate of population growth is seen as too low. Thus in the Russian Federation the main concern is not to increase fertility but to achieve decent living conditions for those already born. The official position of the Russian Federation has been that the low level of fertility is a consequence of persistent socio-economic crisis (ibid.: 18ff.).

An emerging realization is that societies with aging populations require a balanced mix of different policy approaches. Besides policies relating to families and fertility, the mix includes increasing productivity, such as through labour-saving technologies, improving the use of existing labour resources – especially youth, women and older persons – and effecting better coordination of education and labour demand (ibid.: 72). Population enhancement through low mortality, low morbidity, equal opportunity and higher standards of education also constitute population-related goals for aging societies. Another aspect is Demeny's (2003) proposal for a combination of policies supporting moderate migration and a total fertility rate around 1.8 (see Chap. 8). This approach avoids the potential social and economic dislocation consequent upon marked population decline. A slow contraction in labour supply, however, could be sustainable in conjunction with capital deepening and rising labour productivity (McDonald and Moyle 2010: 248). Zero net migration and a moderate rate of population decline could also help to address concerns about resource depletion and global warming if per capita impacts do not increase. However, without positive net migration a population with a sustained TFR of 1.8 would decline by up to 20% in 50 years, depending on its initial age structure. Adverse economic consequences are likely when there is a juxtaposition of pronounced labour force decline and substantial growth in the numbers of older people.

Theories of fertility change in developed countries provide varying perspectives on the potential for policy interventions to promote a general shift from very low fertility. The Second Demographic Transition envisages very little prospect in European countries for higher fertility without thoroughgoing social change. This is outwardly consistent with the long-established belief that governments are unable to cause couples to increase their fertility (United Nations 2004: 25). Further reinforcement comes from the view that population aging is unavoidable. Two or three decades would pass before somewhat higher fertility could affect population aging appreciably, especially when larger cohorts are continuing to reach the older ages. Despite this, the cumulative effect of a small increase in fertility can be substantial in the long run. For example, if Italy had a TFR of 1.3 from 2000 onwards, its percentage aged 65 and over would be about 31 in 2050 compared with 24 if the TFR was about 1.8. Similarly the impact of higher fertility on the number of labour force entries is necessarily delayed, but the longer low fertility continues the more difficult it will be to correct imbalances in the age structure of the labour force.

Compared with the perspective of the second demographic transition, gender equity theory provides a more positive outlook for fertility in developed countries (see Chap. 8). It envisages that the discrimination and burdens that working mothers experience can be addressed through measures to make motherhood and female labour force participation more compatible, such as through affordable child care services, part-time employment and flexible working hours. Some governments, including those of Norway and Sweden, have sought to promote greater gender equity in employment and the family and to enhance the value of children in society (United Nations 2004: 71). For example, in Norway "the Government's 'family friendly' policies, aimed at both men and women, included: 1 year of paid parental leave after each birth, at least one month of which could be taken by the father; arrangements for reduced or flexible work hours for parents; and child allowances" (ibid.: 57). Norway's total fertility rate remained in the range of 1.8–2.0 in the period 1988–2009 (World Bank 2011). Similarly, in the Netherlands, the government has sought a couple-oriented approach – moving away from a focus only on women – and has developed policy measures to "guarantee men's involvement in childcare and childrearing responsibilities." A further example is France's family policies which have been described as "more pronatalist than those of many other developed countries". They include allowances that increase with the number of children and decrease as they get older, together with support for working mothers – such as maternity leave, part-time work and nursery schools providing early childhood nurturing. Associated with this has been one of Europe's highest labour force participation rates for women aged 25–49 (ibid.: 59). Hakim's (2000) preference theory, together with other policy measures that some governments have implemented, further imply that birth rates reasonably close to replacement might be encouraged, not through a single strategy, but through a range of approaches that support family welfare (see Chap. 17). Government influence, therefore, can be a positive force for higher fertility (see Kent and Haub 2005: 19).

In the new demographic situation, international migration is advocated as a means of responding to population aging but, alone, it is not a sufficient response. The most rapidly aging national populations are so large that huge and socially disruptive numbers of immigrants would be required to avert population decline. Thus 'replacement migration' – using migration to counteract an excess of deaths over births – is often unattainable or politically unacceptable. Nevertheless, lower levels of migration can benefit the labour force and, in some situations, reduce the level of aging. In the late 1990s, 61% of population growth in developed regions stemmed from migration, much of it from developing regions. The European Union as a whole has been shifting towards more open borders and, at the same time, better regulation in order to attract selected migrants (United Nations 2004: 71). Also, some countries have successfully operated diverse yet structured immigration policies both to achieve labour force goals, through systematically applied migrant selection criteria, and to meet humanitarian obligations in relation to refugees and family reunion.

In the long run, the aging of immigrants augments the total numbers in the older ages. Thus, for social policies, the consequences of international migration are enduring. They arise initially because of newcomers' housing, employment and welfare needs, and eventually reach renewed prominence for some because of the distinctive issues for aged care potentially arising from the aging of ethnic minorities. Important here are cultural needs, such as for communication in the ethnic language, together with appropriate recognition of dietary preferences, values, traditions and religion. Social inequality also contributes to special needs, since the aged in disadvantaged ethnic groups may be the least educated, the least integrated into the host society, the least proficient in the national language and the most dependent on their families and the welfare system for support.

13.5 Conclusion

The Second World Assembly on Ageing was significant in establishing a wide-ranging and internationally agreed set of responses to individual and population aging. Many issues remain, however, because of the impossibility of implementing a single comprehensive set of strategies in societies with contrasting social and economic characteristics, and because there is considerable latitude for variations within the details of particular policies. Also, the Madrid conference gave little guidance on issues concerning the economic and demographic sustainability of population aging, which underlie the long-term viability of most of the goals that the Assembly identified.

Amid ongoing global economic, social and environmental problems there is mounting questioning of the need for economic growth at all costs (Jackson 2009), as well as of the need for population growth to underpin it (Coleman 2003: 736). The transition to older age structures is significant among global concerns, but national prosperity – based on economic growth, together with maintenance of the size of national labour forces – is likely to be a prerequisite for meeting the challenges that aging will present for health and welfare systems in coming decades. Moreover, although economic considerations are prominent in many national responses to aging, continuing predicaments confronting national economies have made future prosperity uncertain. In such circumstances, population-related interventions are more likely to be delayed or disregarded, even while demographic concerns escalate.

References

Coleman, D. (2003). Population decline. In P. Demeny & G. McNicoll (Eds.), *Encyclopedia of population* (pp. 732–737). New York: Macmillan Reference USA.

Demeny, P. (1988). Demography and the limits to growth. *Population and Development Review*, supplement to Vol. 14 Population and resources in Western intellectual traditions, 213–244.

Demeny, P. (2003). Population policy. In P. Demeny & G. McNicoll (Eds.), *Encyclopedia of population* (pp. 752–763). New York: Macmillan Reference USA.

Hakim, C. (2000). *Work-lifestyle choices in the 21st century: Preference theory*. Oxford: Oxford University Press.

Jackson, T. (2009). *Prosperity without growth: Economics for a finite planet*. London: Earthscan.

Kent, M. M., & Haub, C. (2005). Global demographic divide, *Population Bulletin*, *60*(4), December. Washington: Population Reference Bureau.

Kirakosyan, H. (2002, September 11–13). *From Madrid to Berlin*. In United Nations economic commission for Europe, ministerial conference on ageing, Berlin, Germany: Contribution to the Plenary Session on 13 September 2002 on behalf of the accredited non-governmental organizations, presented by Ms. Hripsime Kirakosyan, Mission Armenia. http://www.sfu.ca/iag/events/ece2.htm. Accessed Apr 2011.

Lutz, W., O'Neill, B., & Scherbov, S. (2004). Europe's population at a turning point. *Vienna Yearbook of Population Research 2004*, 303–312.

McDonald, P., & Moyle, H. (2010). Why do English-speaking countries have relatively high fertility? *Journal of Population Research, 27*(4), 247–273.

Social Integration Branch. (2011). *Mission Statement*. United Nations, Division for Social Policy and Development, Department of Economic and Social Affairs. http://www.un.org/esa/socdev/sib/. Accessed Mar 2011.

UNDESA (United Nations Department of Economic and Social Affairs). (2008). *Guide to the national implementation of the Madrid international plan of action on ageing*. New York: United Nations. http://www.un.org/esa/socdev/ageing/documents/papers/guide.pdf. Accessed Apr 2011.

United Nations. (2002). *Report of the second world assembly on ageing, Madrid, 8–12 April 2002* (A/CONF.197/9), New York: United Nations.

United Nations. (2004). *Policy responses to population decline and ageing* (Population bulletin of the United Nations, Special issue nos. 44/45). New York: Department of Economic and Social Affairs, Population Division.

World Bank. (2011). *World development indicators databank: Fertility rate, total*. http://data.worldbank.org/indicator/SP.DYN.TFRT.IN?cid=GPD_11. Accessed Apr 2011.

Chapter 14
Aging in Asia

> *... the increase in the number of older persons will be greatest in developing countries. This is the most important observation. Over the next 50 years, the older population [aged 60 and over] of the developing world is expected to multiply by four. This is an extraordinary development that bears implications for every community, institution and individual – young and old. Ageing is definitely no longer just a "first world issue". What was a footnote in the twentieth century is on its way to becoming a dominant theme in the twenty-first century.*
>
> (Kofi Annan, in United Nations 2002: 66)

14.1 A Significant Issue?

This chapter discusses population trends in Asia and illustrates contrasts between aging in developing and developed countries. It provides the setting for Chap. 15, which focuses on policy responses to aging in Asia. In 2009, 46% of governments of developing countries identified population aging as a major concern, compared with 79% of the governments of developed countries. Far more prevalent concerns for developing countries were HIV/AIDS, infant, child and maternal mortality, and adolescent fertility (United Nations 2010: 7–10). Their leading age structure issue was the size of the working age population and associated high levels of unemployment. Yet much of the increase in the numbers of older people in the world is occurring in developing countries. Asia's share of the world's aged grew from 44% in 1950 to 51% in 2000 and may reach 61% in 2050 (see Table 1.1). The numbers of people aged 65 and over in Asia could rise from 278 million in 2010 to 906 million in 2050. Much of the increase will occur after 2025, because of the delay before Asia's largest generations reach the older ages. The future pace of change will magnify difficulties in meeting welfare needs in countries that are already unable to provide adequately for their populations.

D.T. Rowland, *Population Aging: The Transformation of Societies*, International Perspectives on Aging 3, DOI 10.1007/978-94-007-4050-1_14,

Despite the magnitude of the numbers, opinion has long been divided over the question of whether aging is a significant issue for Asia. There are strong negative and affirmative arguments about this, with the former predominant until relatively recently. The negative case contends that aging is mainly a matter for developed countries. As the Chief of the Population Estimates and Projections Section of the United Nations Population Division commented: "the current issues of aging relate mostly to the more developed regions" (United Nations 2004: 6). The negative case also emphasizes that the principal indicator of population aging, the percentage in older ages, is low in the developing countries of Asia and will remain so until the second quarter of this century. For example, in many Asian countries less than 5% of the population was aged 65 or more in 2010 and their projected percentages for 2025 are only a little higher (Table 14.1). For decades the relatively low percentage of older people in developing countries was interpreted as indicating that population aging, and with it heightened concern about the welfare of the elderly, was not of particular significance in much of Asia, as well as in Africa and Latin America. Meanwhile, other aspects of population change were often critical considerations.

A further basis for the negative argument is the expectation that the family will continue to assume its 'traditional' responsibilities for the care of dependents, young and old. The notion of the 'demographic dividend', discussed later, also identifies advantages rather than disadvantages in current age structure trends. Thus, rather than viewing aging as a significant issue, some regard concern about aging as a "luxury" for rich countries: rapid population growth, urbanization and economic and social development are more pressing national issues. Poverty, malnutrition and lack of shelter and health care are the most prevalent problems besetting those who grow old in developing countries, but these problems are often indistinguishable from the problems of the population at large (Schade 1982: 99; Heisel 1984: 55). For this reason, there has been resistance to making special provision for older people.

In contrast, the affirmative case focuses especially on the total numbers of the aged and their vulnerability, arguing that relatively low percentages in older ages disguise problems and their severity. As emphasized in the quotation at the start of this chapter, the predominant trend in aging in developing countries is the inevitability of great increases in the numbers of older people. The main issue is the already precarious existence of expanding multitudes of the aged. Many of the personal predicaments of individual older people remain the same whatever the percentages. This is because of widespread poverty and disease in conjunction with inadequate pension systems and low resource levels per capita. Also, although the growing numbers of the aged is the dominant issue, demographic aging itself is occurring much more rapidly in parts of Asia than in it did in Western countries. While the greatest impacts of growing numbers and percentages lie ahead, adequate responses require a long lead time to secure real benefits from many initiatives – including promoting health, improving income support and fostering attitudinal changes that combat ageism and facilitate social participation.

Affirmative arguments also include skepticism about the efficacy of traditional family support: (i) because urbanization and other social changes are weakening the family's ability to provide care; (ii) because the proportion of families with aged

Table 14.1 Population growth and aging in Asia[a]

Region/country	Total population data					Percentage aged 65 and over			Numbers aged 65 and over (millions)			Dependency ratios 2025	
	Total fertility rate 2005–2010	Life expectancy at birth 2005–2010 M	Life expectancy at birth 2005–2010 F	Annual population growth rate % 2005–2010	Doubling time in years	2010	2025	2050	2010	2025	2050	Total	Aged
ASIA	2.35	67	71	1.14	61	6.7	9.9	17.3	278.3	472.4	906.1	48	15
Eastern Asia	1.72	72	76	0.56	124	9.5	14.7	24.5	148.9	244.4	392.2	47	22
China	1.77	71	75	0.63	110	8.2	13.4	23.3	111.4	194.2	330.6	46	19
Hong Kong	1.02	79	85	0.54	128	12.9	22.1	32.6	0.9	1.8	2.8	52	34
Japan	1.27	79	86	–0.07	–	22.6	29.7	37.8	28.7	35.8	38.5	69	50
North Korea	1.86	65	69	0.39	178	9.8	11.0	18.1	2.4	2.8	4.4	43	16
South Korea	1.22	76	82	0.39	178	11.0	19.3	34.2	5.3	9.6	15.1	48	28
South-central Asia	2.82	63	66	1.51	46	4.7	6.9	13.3	84.0	147.3	331.6	49	10
Afghanistan	6.63	44	44	3.45	20	2.2	2.4	3.6	0.7	1.1	2.7	83	4
Bangladesh	2.36	65	67	1.42	49	4.0	6.1	14.9	6.5	11.9	33.2	44	9
India	2.76	62	65	1.43	48	4.9	7.3	13.7	59.7	105.0	221.8	47	11
Iran	1.83	70	73	1.18	59	4.8	7.4	19.7	3.6	6.4	19.1	41	10
Nepal	2.94	66	67	1.85	37	4.1	5.2	10.6	1.2	2.0	5.2	52	8
Pakistan	4.00	66	67	2.16	32	4.1	5.1	10.0	7.5	12.7	33.5	58	8
Sri Lanka	2.33	70	78	0.88	79	7.7	13.9	21.4	1.6	3.1	4.6	54	21
Uzbekistan	2.29	65	71	1.09	64	4.4	6.9	14.8	1.2	2.2	5.4	46	10
South-eastern Asia	2.32	68	72	1.24	56	5.9	9.0	17.3	34.5	61.4	132.3	46	13
Indonesia	2.19	69	73	1.18	59	6.1	9.0	18.6	14.1	23.6	53.6	43	13
Malaysia	2.58	72	77	1.71	41	4.8	8.7	16.3	1.4	2.9	6.5	48	13
Myanmar	2.32	59	63	0.87	80	5.5	8.6	17.5	2.8	5.0	11.1	46	13
Philippines	3.11	70	74	1.82	38	4.3	6.6	12.7	4.0	7.7	18.5	53	10
Singapore	1.27	78	83	2.51	28	10.2	22.9	32.6	0.5	1.2	1.7	55	35

(continued)

Table 14.1 (continued)

Region/country	Total population data					Percentage aged 65 and over			Numbers aged 65 and over (millions)			Dependency ratios 2025	
	Total fertility rate 2005–2010	Life expectancy at birth 2005–2010 M	Life expectancy at birth 2005–2010 F	Annual population growth rate % 2005–2010	Doubling time in years	2010	2025	2050	2010	2025	2050	Total	Aged
Thailand	1.81	66	72	0.65	107	7.7	12.9	20.2	5.3	9.4	14.8	48	19
Viet Nam	2.08	72	76	1.15	60	6.3	9.8	20.0	5.7	10.0	22.4	44	14
Western Asia	2.95	69	74	1.95	36	4.7	6.6	13.4	10.9	19.3	50.0	50	10
Iraq	4.11	64	72	2.17	32	3.2	4.1	8.9	1.0	1.8	5.7	58	7
Saudi Arabia	3.17	71	75	2.12	33	3.0	5.2	12.1	0.8	1.8	5.9	46	8
Syria	3.29	72	76	3.26	21	3.2	4.8	13.4	0.7	1.4	5.0	51	7
Turkey	2.13	69	74	1.24	56	6.0	8.8	18.4	4.5	7.7	17.9	44	13
Yemen	5.30	61	64	2.86	24	2.4	3.1	6.4	0.6	1.1	3.4	68	5

Source: United Nations (2009)

[a] The table presents medium variant projections for countries with populations greater than 20 million in 2010, together with Hong Kong and Singapore

relatives is increasing while their status is declining; and (iii) because higher proportions of people are surviving to ages where infirmities are most prevalent. In developing countries, family care has an expanded role because of the scarcity of state-funded pensions and services. Although family support for the aged is a long-established expectation, low rates of survival in the past meant that only a small proportion of families included older relatives. The widening prevalence during the late twentieth century of families with at least three surviving generations became a major new feature. Families now face greater responsibilities in caring for dependents. Higher labour force participation of women and the separation of relatives through migration also make family support more difficult.

Asia is a diverse region in which population trends and their implications vary considerably between countries, as well as within them. The percentage aged 65 and over is widely used as the only social indicator of the extent of welfare needs among the aged and the 'burden' of aging on economies and the societies. However, it is not sufficient in itself. Other indicators are needed to reveal the circumstances of the aged and consequences for each society – such as measures of income, health and social inequality. Even where the percentage in older ages is low, welfare needs among the elderly can be great, but neglected. Progress depends on integrating concerns for young and old, creating a society in which no age group is marginalized.

14.2 Recognition of Aging

Recognition of aging as an issue for developing countries was delayed because of its identification solely with increases in the proportion in older groups. There seemed little justification for giving special attention to the welfare of the aged because children and their parents were more numerous and all age groups had pressing needs. One of the first to take issue with this position was the Duke University demographer George Myers (1982: 3) who criticized the emphasis demographic theory placed on the proportion of the population in older ages, to the exclusion of their numbers, and called for policies to give proper recognition to the well-being of all segments of the population. Although all age groups are equally deserving, their needs differ. For older people, distinctive considerations are age-related vulnerability, the effects of family change on their roles and means of support, together with their potential, beyond traditional and stereotypical expectations, to contribute to society. The relatively recent concept of 'a society for all ages' acknowledges the importance of all age groups as well as their different needs and capacities.

By the mid-1980s there was wider acknowledgement that growth in the numbers of the aged was important (Heisel 1984: 50). As El-Badry (1988: 396) commented:

> The static proportion of the aged poorly presents the magnitude of their problems in developing countries because the key issue is the balance between the growth in the size of that group and the speed at which the basic services needed can be provided in competition with all other demands from the whole population.

Furthermore, even in the mid-1980s observers were commenting that: "the pace of aging in less developed countries seems well ahead of institutional preparedness to meet health and social welfare needs" (Treas and Logue 1986: 646). The 2002 Second World Assembly on Ageing gave recognition to the increasing significance of aging in developing countries, whereas the First World Assembly on Ageing, held in Vienna in 1982, had focused on older persons in developed nations (Andrews 2005: xvii). Other more recent conferences on aging in developing countries, together with the further refinement of policy documents, attest to the current importance of this issue.

14.3 Prospects

Over their transition from high to low birth and death rates, the now more developed countries mostly doubled or trebled their total populations (McNicoll 1984: 181). By contrast many developing countries could increase their numbers six to tenfold during this transition. Also, while the numbers of the aged in more developed regions are approximately doubling in 50 years, in less developed regions they are quadrupling. The number of people aged 65 and over in less developed regions probably exceeded those in more developed regions as early as 1950. By 2000, the numbers were 46% greater than in the more developed regions, and could be 260% greater by 2050. Much of the less developed regions' population growth for decades ahead is inevitable because of the in-built momentum in their current age structures. The highest growth rates will be in the older ages. This reflects the aging of the large generations born in the 1950s and 1960s when death rates were falling while birth rates remained high. There is a parallel here with the baby boom generations born in developed countries after the Second World War, as major social and economic consequences have arisen from the shock wave due to their larger numbers advancing through successive stages of life. In comparison the Asian shock wave is a tsunami. By the year 2000, 15 countries had aged populations numbering 5 million or more, compared with a prospective 28 countries in 2025 and 42 in 2050. In 2000 only 5 of these were developing countries, but they will comprise the majority in coming years (Table 14.2).

Containment of the size of the aged population in developing countries will depend on minimizing the duration of above replacement fertility. Continuing high birth rates in some countries, in conjunction with lower death rates, are producing rapid expansion in the numbers of children. This is perpetuating a cycle of self-reinforcing growth as ever greater numbers reach the reproductive ages and ultimately the older ages. Model populations in Table 14.3 illustrate the long-run implications of different fertility and mortality rates for population growth in Asia's developing countries. In the table, female life expectancies of 65–75 cover the current range in different regions of Asia, while total fertility rates of 2–4 similarly encompass much of the range for Asia's developing countries, apart from a few high figures such as the TFR of around six children per woman in Afghanistan.

Table 14.2 Countries with the world's largest populations aged 65 years and over, 1950–2050 (millions)

1950		1975		2000		2025		2050		Row
China	24	China	41	China	86	China	194	China	331	1
United States	13	United States	23	India	44	India	105	India	222	2
India	12	India	21	United States	36	United States	65	United States	87	3
Germany	7	Russia	12	Japan	22	Japan	36	Indonesia	54	4
Russia	6	Germany	12	Russia	18	Brazil	24	Brazil	49	5
United Kingdom	5	Japan	9	Germany	13	Indonesia	24	Japan	38	6
		United Kingdom	8	Italy	11	Russia	23	Pakistan	34	7
		France	7	Indonesia	10	Germany	20	Bangladesh	33	8
		Italy	7	France	10	France	15	Mexico	29	9
		Ukraine	5	Brazil	10	Italy	15	Russia	27	10
				United Kingdom	9	Mexico	13	Germany	23	11
				Ukraine	7	United Kingdom	13	Viet Nam	22	12
				Spain	7	Pakistan	13	Iran	19	13
				Pakistan	5	Bangladesh	12	Italy	19	14
				Mexico	5	Spain	10	Philippines	19	15
						Viet Nam	10	France	18	16

(continued)

Table 14.2 (continued)

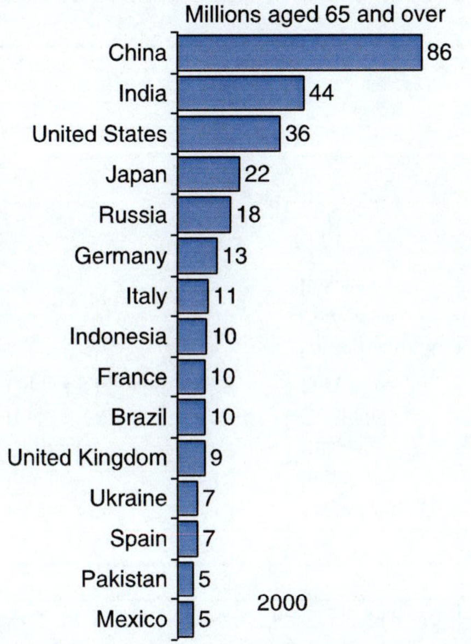

1950	1975	2000	2025		2050	Row
		South Korea	10	Nigeria	18	17
		Thailand	9	Turkey	18	18
		Canada	8	Egypt	17	19
		Poland	8	United Kingdom	17	20
		Ukraine	8	Spain	16	21
		Philippines	8	Korea	15	22
		Turkey	8	Thailand	15	23
		Nigeria	8	Canada	11	24
		Egypt	7	Myanmar	11	25
		Iran	6	Colombia	11	26
		Argentina	6	Ethiopia	11	27
		Colombia	5	Argentina	10	28
				Poland	10	29
				Algeria	9	30
				Ukraine	9	31
				Morocco	7	32
				Congo	7	33
				Venezuela	7	34
				Australia	7	35
				Peru	7	36
				Sudan	7	37
				Malaysia	6	38
				Saudi Arabia	6	39
				Iraq	6	40
				Tanzania	6	41
				Uzbekistan	5	42

Source: United Nations (2009)

Each model has a constant age structure and a constant growth rate. Therefore the models with replacement level fertility have, from the outset, zero growth together with zero momentum in their age structures. In real populations, zero growth normally cannot occur until many years after replacement fertility is reached, because of inherent population momentum.

A key indicator of the pace of change in Table 14.3 is the doubling time of a population. This is only 30–36 years for a total fertility rate of 4, but slows substantially for each lower level of fertility. Smaller variations in doubling times arise from differences in life expectancies. Only as the birth rate approaches replacement level (which varies between total fertility rates of 2.3 and 2.1, depending on mortality levels) does population growth become more manageable. The statistics in the table also illustrate that the future numbers of the aged are closely linked to birth rates and that low percentages 65 or older belie the extent of population growth occurring both in younger and older age groups. With relatively high life expectancies in most of Asia, birth rates of 2.5 or more will result in very high increases. The statistics indicate that aging in Asia is a matter not only for responsive social welfare policies but also for proactive policies that seek to restrain population growth.

United Nations (2009) medium variant projections for Asia assume that policies and social and economic changes will ameliorate population trends. Much of the future growth of Asia's population is projected to occur in age groups 15–64, reflecting the assumption that the total fertility rate for the whole region will converge towards replacement, falling gradually from 2.4 in 2005–2010 to just below 2.0 in the 2030s and 2040s. This would maintain relatively low total dependency levels, supporting the view that trends in dependency ratios are "neutral to favourable for development in all regions" (United Nations 1990: 48). Yet the real concern remains the welfare of Asia's vast and expanding numbers of older people. Also, most of Eastern Asia, together with Iran, Singapore and Thailand, are already on a trajectory to heightened levels of aging associated with below replacement fertility (Table 14.1).

As shown in Chap. 3, Asia's age structure evolved from 'very young' to 'young' in the last quarter of the twentieth century and is projected to be 'mature' by 2025. Although age structure changes after 2025 are more uncertain, United Nations medium variant figures for the whole of Asia are consistent with an 'aging' profile emerging by mid-century. Figure 14.1 illustrates these changes, including the narrowing of the base of Asia's age structure by 2050. The latter augurs labour force decline in the second half of the century, whereas, in the first half, enormous growth in the labour force ages will accompany the rapid expansion in the size of Asia's aged population. About 80% of India's population growth 2000–2050, for example, will be in ages 15–64. The regions of Asia exhibit a similar pattern of change towards older age structures over time, except for Eastern Asia (i.e. China, Hong Kong, Macao, Japan, North and South Korea, Mongolia and Taiwan). In that region, rapid aging due to low fertility may produce an 'aging' profile in 2025 and a 'very old' profile in 2050.

At the country level there is greater variation between Asia's age structures, as illustrated in Fig. 14.2. The region has the world's oldest population – Japan – but

Table 14.3 Stable population models illustrating prospective changes in Asian developing countries

Total fertility rate per woman (replacement TFR in bold)	Intrinsic growth rate %	Doubling time in years	% aged 65 and over	% aged 75 and over	Total dependency ratio	Aged dependency ratio	% growth in the total population and the aged population after 50 years
Life expectancy 65 (females), 61.2 (males)							
4.0	1.9	35.6	5.7	1.8	69.3	9.7	165.0
3.5	1.5	47.2	7.1	2.3	63.7	11.6	108.3
3.0	0.9	74.6	8.9	3.0	58.8	14.1	59.1
2.5	0.3	244.7	11.5	4.1	54.6	17.8	15.2
2.3	0.0	–	12.8	4.7	53.3	19.6	0.0
Life expectancy 70 (females), 66 (males)							
4.0	2.1	32.6	5.9	1.9	70.7	10.1	189.5
3.5	1.6	42.1	7.3	2.5	65.2	12.0	127.9
3.0	1.1	63.1	9.2	3.3	60.1	14.7	73.2
2.5	0.5	147.5	11.8	4.4	56.0	18.4	26.5
2.2	0.0	–	14.1	5.4	54.3	21.8	0.0
Life expectancy 75 (females), 71.2 (males)							
4.0	2.2	30.8	6.4	2.3	72.1	11.1	207.7
3.5	1.8	39.0	7.9	2.9	66.8	13.2	142.9
3.0	1.2	56.1	10.0	3.8	62.0	16.1	85.4
2.5	0.6	117.7	12.9	5.1	58.2	20.4	34.2
2.1	0.0	–	16.1	6.6	56.6	25.1	0.0

Sources: West Model Life Tables: Coale and Demeny (1983), Coale and Guo (1990), Rowland (2003: 316–319)
Note: Table calculated from stable population models with specified TFRs and life expectancies

Fig. 14.1 Asia, age structures 1950, 2000 and 2050 (Source: United Nations 2009, estimates and medium variant projections)

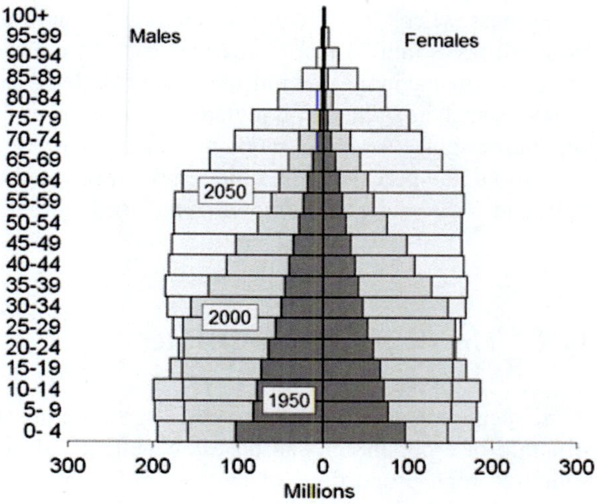

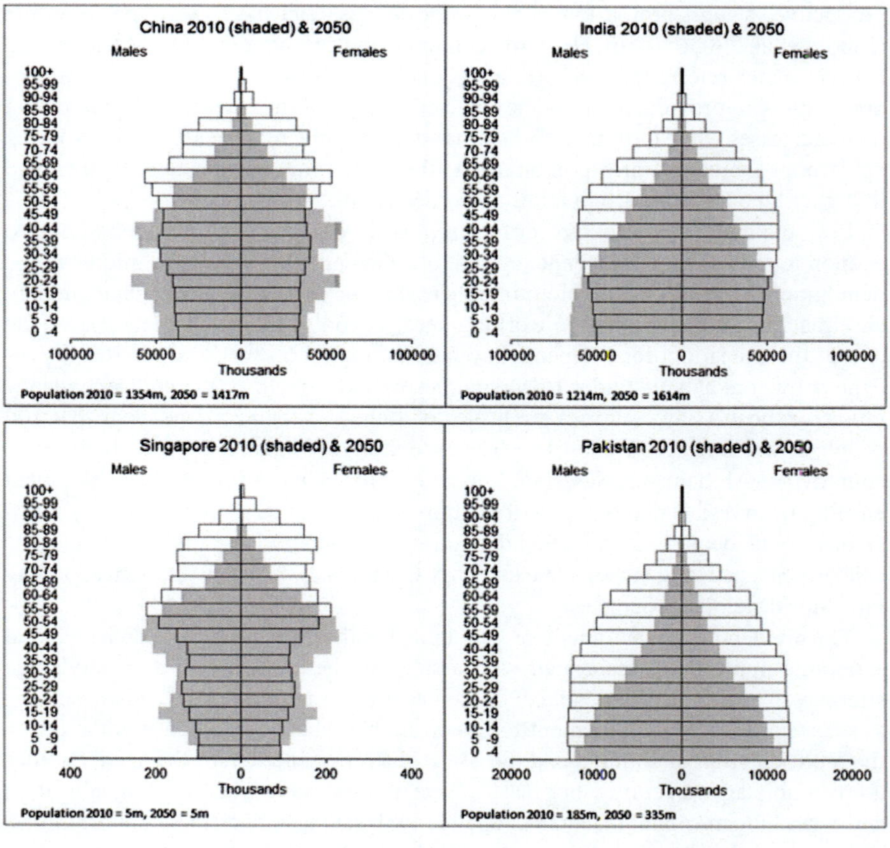

Fig. 14.2 Age structures of selected Asian country populations 2010 and 2050 (Source: United Nations 2009, medium variant projections)

also others where low fertility is leading to rapid aging, notably in Singapore, Hong Kong, South Korea, China and Iran. In contrast, high fertility in Iraq and Pakistan is maintaining youthful age structures. Between these two sets of extremes are countries such as India and Bangladesh where lower fertility is leading to moderate population aging. Nevertheless change, albeit at different scales, is pervasive. In Asia as a whole, the percentage in older ages is projected to more than double between 2010 and 2050, while the numbers of the aged more than treble (Table 14.1).

14.4 The Demographic Dividend

By no means all current trends in Asia's age structures are disadvantageous. The age structure of a population can have a beneficial impact on economic performance. Although high proportions of children or elderly people in a population tend to restrain the rate of economic growth, because the young and the old are usually less productive, a high proportion in the working ages can have the opposite effect (Bloom et al. 2003: xi–xii). Harvard economist and demographer David Bloom and his colleagues referred to the latter as producing a 'demographic dividend' of economic growth, provided that policies take advantage of this situation. The potential dividend arises when birth rates fall substantially and the relative size of the working age groups increases correspondingly. At the same time, smaller family sizes lead to higher labour force participation of women (ibid.: 39).

Low dependency ratios are crude indicators of age structures conducive to economic growth. Child dependency and total dependency peak in the middle of the demographic transition, at which time there may be around 90 dependants, mainly children, per 100 people in the working ages, 15–64 years (see Table 1.3). By the end of the transition, total dependency falls to around 60 workers per 100 dependants, of whom half are under 15 and half are 65 or more. In 2025, total dependency ratios for most Asian countries are projected to be less than 55 dependents per 100 people of working age (Table 14.1). The exceptions are Japan and the high birth rate countries of Afghanistan, Pakistan, Iraq and Yemen. Thus a favorable development ensuing from Asia's demographic transition is the fall in its total dependency ratios as birth rates decline. Aged dependency ratios in Asia will gradually increase as child dependency falls, but for some time they will remain well below the corresponding ratios for developed countries.

The dividend is not automatic or guaranteed; rather, it is no more than a potential economic opportunity associated with a stage in the evolution of a society's age structure. To reap the dividend, or 'bonus' as it is sometimes called, it is necessary to secure productive employment for the available labour force between the fall in the representation of children and the rise in the representation of the aged. Besides a favourable age structure other factors have an important role including education and other human capital resources, together with savings, investment opportunities, and access to global markets (Bloom et al. 2003: 39–42). Without these other advantages, high unemployment is more likely than the dividend (Magnus 2009: 55). In Indonesia and the Philippines unemployment rates in the 3 years to 2010 were in the range of 7–8% for both sexes, compared with 4–5% in Japan, and 3–4% in South Korea and Malaysia (ILO 2011). Even with low unemployment the dividend is not inevitable if labour productivity is low and individual earnings are insufficient to prevent poverty. In India about half of all employed people in 2005 had less than US$1 per day to live on, as did more than a quarter in Indonesia and the Philippines (United Nations 2011).

As examples of successful exploitation of the dividend, Bloom and his colleagues refer to Eastern Asian economies where, between 1965 and 1990 the region's working age population grew nearly four times faster than the dependent population. Their estimates suggest that the demographic dividend accounted for between 25% and 40% of Eastern Asia's 'economic miracle' (ibid.: 45). In Japan, however, population aging has curtailed the influence of the demographic dividend and there is little international migration supplementing the numbers in the working ages. Latin American countries have had age structures favorable to the dividend, but "weak governance and a lack of openness to trade" slowed the potential growth (ibid.: 58). Southeast Asia has recently begun to benefit from the demographic dividend, while China's is "pretty much exhausted": its working age population is expected to decline slowly 2010–2030 and more rapidly thereafter as the aging of its population gathers pace (Magnus 2009: 158ff.). In comparison, the United States has been benefitting from the demographic dividend, because of its relatively low dependency ratio and high level of immigration, while Europe's demographic dividend "is about to expire" (Bloom et al. 2003: 43ff).

The potential for the 'dividend' is present fairly late in the demographic transition – in the shift from young to mature age structures – as well as in the early years of the second demographic transition. The potential for a demographic dividend wanes when the relatively large cohorts in the working ages begin to reach older ages. The only subsequent type of age profile that retains a fairly high proportion in the working ages is one which approximates the theoretical end-point of the demographic transition, namely a rectangular age structure with near-zero growth and near-replacement fertility. It minimizes overall dependency, retains relatively large numbers in the main working ages and avoids endless population growth or decline.

Among the international agencies using the demographic dividend concept are the World Health Organization and the United Nations Population Fund (UNFPA), for instance to demonstrate the economic advantages of investing in young people (Hendrixson 2007: 18). Advocates of family planning in developing countries have also emphasized the dividend. However, this has evoked concerns that "It mobilizes family planning as a means to achieve an economic end, rather than to promote information on and access to contraceptive methods as a fundamental human right" (ibid.: 18). Also, references to population aging as an 'onus', diminishing the demographic 'bonus', risk portraying consequences of aging solely in negative economic terms, rather than recognizing the positive roles of older people and the potential to foster these further.

14.5 The Status of the Aged

Besides realization of the demographic dividend, a vital aspect of future social and economic change in developing countries is an improvement in the status of older people. Little is known about shifts in the political power and status of the elderly over time, but such changes are thought to accompany social and economic modernization manifested in the demographic transition and the transformation of national age structures. A perceived later commencement of old age is sometimes associated with modernization because of increased longevity and improvements in health. An influential hypothesis about changes in the status of the aged is that the allocation of resources and honour to older persons declines as modernization occurs (Cowgill 1972; Palmore 2005):

> Modernization is declared to be associated with later onset of old age, increased use of chronological criteria, increased longevity, an aging population, increased proportions of females and widows in the population, increased proportions of grandparents, lower status of the aged, decline in leadership roles of the aged, decline in power and influence of the aged, increased ambiguity of the role of widows and an increase in the extent of disengagement of older people from community life. (Cowgill 1974, cited by Driedger and Chappell 1987: 19).

Proposed mechanisms responsible for loss of status are: (i) the break up of the extended family as the household unit together with increases in the spatial separation of generations, which reduces family interaction and interdependence; (ii) rising standards of education and shifts in the knowledge base of society leading to a devaluation of the knowledge and experience of the aged; and (iii) a decline in death rates leading to older people becoming more numerous. Eventually the status of the aged may rise as the impact of some changes recede, for example if generations become more similar again in terms of literacy, educational and occupational backgrounds and family values. Trela and Sokolovsky (1979: 121–122) suggested that the status of the aged is maximized under a number of conditions, including when useful and valued functions are continued as long as possible, and when the extended family is a viable residential or economic unit.

Early testing of modernization theory suggested that the aged lose status during initial rapid changes – for instance as the young become literate – but later, as change becomes slower, their relative status improves (Maddox and Wiley 1976: 10). Exceptions include the apparent continuation of high status for the aged in Japan, Ireland, and Russia, reflecting adherence to earlier cultural values (McPherson 1990: 43). A comparison of 31 countries concluded that it is important to distinguish between short-run and long-run outcomes of social change (Palmore and Manton 1974). Studies of traditional societies have found that in cultures where deference or honour is paid to the old, they nonetheless may have limited access to food or other material resources (Neugarten and Hagestad 1976: 38). Also, since the elderly are a diverse group, not everyone necessarily experiences a change of status as they grow older (McPherson 1990: 44). Importantly, their standing in their own family and community may differ from that in the broader society.

Much of the research on aging and modernization has focused on ethnic minorities in developed countries. Hence findings and criticisms of the theory are not necessarily transferable to developing countries. Rosenthal (1983) and others have presented critiques of modernization theory, and Driedger and Chappell (1987: 23–27) and Markides and Mindel (1987: 29) have summarized this literature. A firm defence of the theory appeared in its originators' later work: Cowgill (1986) and Holmes (1987). Key criticisms of modernization theory are that it fails to differentiate between ethnic groups and assumes that modern families are unsupportive of the aged (Ujimoto 1983: v). More recent commentaries have described modernization theory as an oversimplification which ignores cross-cultural differences in the values and belief systems that influence responses to socio-economic development (Löckenhoff et al. 2009: 941; Fry 2009: 518). The well-being of Asia's future aged will depend considerably on the extent to which their society accords them status and resources equal to that of other age groups.

14.6 Conclusion

Core issues for Asia are the pace of growth and change in its total population and in its aged population. The peak in the growth rate of the aged population will occur about 60 years after the late 1960s peak in the growth rate of the total population. At mid-century, more than 60% of Asia's elderly will be in China and India. Although older age structures bring to the fore the need for greater attention to the circumstances of the elderly, the same is true in younger age structures when the numbers 65 and over are growing rapidly. Asia's low overall level of demographic aging long encouraged dismissive attitudes to it, with the unintended consequence of marginalizing the needs of older people. More recently, opposite views have brought forth growing national and international efforts to address this situation. Asian governments have adopted their own major policy documents setting agendas for reform, one of which predated the 2002 Madrid Plan of Action. These initiatives are the subject of the next chapter.

References

Andrews, G. R. (2005). Foreword. In M. L. Johnson (Ed.), *The Cambridge handbook of age and ageing* (pp. xvii–xix). Cambridge: Cambridge University Press.

Bloom, D. E., Canning, D., & Sevilla, J. (2003). *The demographic dividend: A new perspective on the economic consequences of population change*. Santa Monica: Rand.

Coale, A. J., & Demeny, P. (1983). *Regional model life tables and stable populations*. New York: Academic Press.

Coale, A. J., & Guo, G. (1990). New regional model life tables at high expectation of life. *Population Index, 56*, 26–41.

Cowgill, D. (1972). A theory of aging in cross-cultural perspective. In D. Cowgill & L. Holmes (Eds.), *Aging and modernization*. New York: Appleton.

Cowgill, D. (1974). Ageing and modernization: A revision of the theory. In J. Gubrium (Ed.), *Later life: Communities and environmental policy*. Springfield: Charles C. Thomas.

Cowgill, D. O. (1986). *Aging around the world*. Belmont: Wadsworth Publishing Company.

Driedger, L., & Chappell, N. (1987). *Aging and ethnicity: Toward an interface*. Toronto: Butterworths.

El-Badry, M. A. (1988). Aging in developing countries: One more population problem? In United Nations (Ed.), *Economic and social implications of population aging* (pp. 389–398). New York: United Nations, Department of International Economic and Social Affairs, ST/ESA/SER.R/85.

Fry, C. L. (2009). Out of the armchair and off the veranda: Anthropological theories and the experiences of aging. In V. L. Bengtson, D. Gans, N. Putney, & M. Silverstein (Eds.), *Handbook of theories of aging* (pp. 517–528). New York: Springer Publishing Company.

Heisel, M. A. (1984). Aging in the context of population policies in developing countries. *Population Bulletin of the United Nations*, (17), 49–63.

Hendrixson, A. (2007, Spring). What's wrong with the 'demographic dividend' concept? *Population Policy Initiative (Babies, burdens and threats: Current faces of population control),* 17–20.

Holmes, L. D. (1987). Cultural values and cultural change: A cross-cultural analysis of aging and age status. *Journal of Cross-Cultural Gerontology, 2*, 195–200.

ILO. (2011). Short term indicators of the labour market. *LABORSTA Internet*. ILO Department of Statistics. http://laborsta.ilo.org/sti/sti_E.html. Accessed May 2011.

Löckenhoff, C. E., et al. (2009). Perceptions of aging across 26 cultures and their culture-level associates. *Psychology and Aging, 24*(4), 941–954.

Maddox, G. L., & Wiley, J. (1976). Scope, concepts and methods in the study of aging. In R. H. Binstock & E. Shanas (Eds.), *Handbook of aging and the social sciences* (pp. 3–34). New York: Van Nostrand Reinhold.

Magnus, G. (2009). *The age of aging: How demographics are changing the global economy and our world*. Singapore: Wiley (Asia).

Markides, K. S., & Mindel, C. H. (1987). *Aging and ethnicity*. Newbury Park: Sage Publications.

McNicoll, G. (1984). Consequences of rapid population growth: Overview and assessment. *Population and Development Review, 10*(2), 177–240.

McPherson, B. D. (1990). *Aging as a social process*. Toronto: Butterworths.

Myers, G. C. (1982). The aging of populations. In R. H. Binstock, W. S. Chow, & J. H. Schultz (Eds.), *International perspectives on aging: Population and policy challenges* (Policy development studies, no. 7, pp. 1–37). New York: United Nations Fund for Population Activities.

Neugarten, B. L., & Hagestad, G. O. (1976). Age and the life course. In R. H. Binstock & E. Shanas (Eds.), *Handbook of aging and the social sciences* (pp. 35–55). New York: Van Nostrand Reinhold.

Palmore, E. B. (2005). Modernization theory. In E. B. Palmore, L. G. Branch, & D. K. Harris (Eds.), *Encyclopedia of ageism* (pp. 231–232). New York: Haworth Press.

Palmore, E. B., & Manton, K. (1974). Modernization and status of the aged: International correlations. *Journal of Gerontology, 29*(2), 205–210.

Rosenthal, C. J. (1983). Aging, ethnicity and the family: Beyond the modernization thesis. *Canadian Ethnic Studies, 15*(3), 1–16.

Rowland, D. T. (2003). *Demographic methods and concepts*. Oxford: Oxford University Press.

Schade, B. (1982). Aging and old age in developing countries. In H. Thomae & G. Maddox (Eds.), *New perspectives on old age: A message to decision makers* (pp. 98–112). New York: Springer.

Treas, J., & Logue, B. (1986). Economic development and the older population. *Population and Development Review, 12*, 645–673.

Trela, J. E., & Sokolovsky, J. H. (1979). Culture, ethnicity, and policy for the aged. In D. E. Gelfand & A. J. Kutzik (Eds.), *Ethnicity and aging: Theory, research and policy* (pp. 117–136). New York: Springer.

Ujimoto, V. K. (1983). Introduction: Ethnicity and aging in Canada. *Canadian Ethnic Studies, 15*(3), 3–8.

United Nations. (1990). *World population monitoring 1989*. Population Studies No. 113, New York: Department of International Economic and Social Affairs.

United Nations. (2002, April 8–12). *Report of the second world assembly on ageing*, Madrid. New York: United Nations.

United Nations. (2004). *Policy responses to population decline and ageing* (Population bulletin of the United Nations, special issue nos. 44/45). New York: Department of Economic and Social Affairs, Population Division.

United Nations. (2009). *World population prospects: The 2008 revision population database*. New York: Population Division of the Department of Economic and Social Affairs. http://esa.un.org/UNPP/index.asp?panel=2. Accessed Apr 2010.

United Nations. (2010). *World population policies 2009*. New York: Department of Economic and Social Affairs, Population Division.

United Nations. (2011). *Millenium development goals indicators: The official United Nations site for the MDG indicators*. United Nations, Department of Economic and Social Affairs. http://mdgs.un.org/unsd/mdg/Host.aspx?Content=Indicators/OfficialList.htm. Accessed May 2011.

Chapter 15
Policy Responses in Asia

It is essential to integrate the evolving process of global ageing within the larger process of development. Policies on ageing deserve close examination from the developmental perspective of a broader life course and a society-wide view, taking into account recent global initiatives and the guiding principles set down by major United Nations conferences and summits.

(United Nations 2002: 6)

15.1 The Macau Plan

This chapter first discusses Asia's main international policy documents concerned with responses to aging. Later sections highlight major concerns about poverty, health and the vulnerability of the family. The chapter aims to clarify key issues in relation to the aged in Asia's developing countries. It also illustrates the importance of international inquiry, discussion and action in formulating and implementing policies. In preparation for the 1999 International Year of Older Persons, the Economic and Social Commission for Asia and the Pacific (ESCAP), a regional arm of the United Nations, convened a meeting in Macau in 1998 which developed a *Plan of Action on Ageing for Asia and the Pacific* – 4 years before the Second World Assembly on Ageing (ESCAP 1998a). The Macau Plan, which governments in the region endorsed in 1999, gave particular attention to the needs of developing countries in the region, recognizing that their priority was overall social and economic development, including the provision of basic social services for all: "Hunger, poverty, ill health, social exclusion, unemployment, and limited access to education and basic amenities are critical national areas of concern requiring corrective action and competing for scarce resources" (ibid.: 2).

Thus the Macau Plan was one of the first international policy documents to address the question of aging and development. Issues concerning aging were

D.T. Rowland, *Population Aging: The Transformation of Societies*, International Perspectives on Aging 3, DOI 10.1007/978-94-007-4050-1_15,

viewed within a broad developmental context, for example recognizing the aged as contributors to national development as well as a group with particular needs. Similarly, the Plan acknowledged that social and economic development was transforming national institutions and values that impact on older people, such as through changes in the nature of families and communities. Although later formulations have refined or revised parts of the Plan, it was a pioneering endeavour declaring a vision of promoting the well-being and participation of older people, with particular concern for women, the poor and people in rural areas (ibid.: 10). The needs of widows and others without family support underlay the Plan's focus on women. Some of the Plan's proposals, however, were more relevant to an urban middle class than to the urban and rural poor. While noting that governmental interest in aging and the aged varied between countries, the Macau Plan identified four issues of broad relevance for immediate action namely:

> (i) understanding of the issues and implications of population ageing on society; (ii) preparation of the population for an ageing process that is productive and fulfilling; (iii) development of a service infrastructure and environment based on traditional and modern institutions that will be able to meet present and future needs; and (iv) delivery of essential services needed by the growing number of older persons. (ibid.: 3–4).

Discussion of these ranged across a number of major subject areas including negative stereotypes, income, employment, the family and service provision. Dispelling negative perceptions was acknowledged as a prerequisite for combating marginalization of older people and fostering positive, productive roles for them in the workplace, the community and the family. Lifelong education and equal opportunity were endorsed as conducive to productive aging. The Macau Plan's discussion of income in later life mentioned the importance of life-long saving, employment opportunities for older workers and the role of the family in providing financial security. It advocated the abandonment of a rigid retirement age and legal protection to enable capable older persons to remain economically active (ibid.: 9).

The Plan emphasized the roles of the family as the basic social unit in which support is given and received and as the economic unit engaging in productive activities, especially in rural areas. It noted that in most of the region's plans for population aging the family remains the "front line institution", although there was also special mention of the contribution of older people to the welfare of the family and younger generations. The Plan recognized that the family's capacity for aged care was diminishing because of smaller family sizes, higher labour force participation of women and migration. In response to this, governments needed to enhance the care-giving capacity of the family through promotion of co-residence and provision of home-nursing, respite care and counseling. For those without family support it was necessary for governments to make special provision (ibid.: 5).

The Macau Plan's approach to health and health services was similar to that advocated for developed countries, emphasizing primary prevention, healthy lifestyles and maintenance of physical and social activity. Educating the population on good nutrition and healthy aging, and disseminating information about common illnesses such as diabetes, cataract and arthritis, were seen as affordable strategies that could be integrated with existing community health programs. These approaches, rather than high cost medical technologies, were thought to have most significance for the

Table 15.1 Major directions in the Macao, Madrid and Shanghai policy documents

Macao 1999	Madrid 2002	Shanghai 2002
Understanding of the issues and implications of population aging on society.	Older persons and development.	Older persons and development.
Preparation of the population for an aging process that is productive and fulfilling.	Advancing health and well-being into old age.	Advancing health and well-being into old age.
Development of a service infrastructure and environment based on traditional and modern institutions that will be able to meet present and future needs.	Ensuring enabling and supportive environments.	Ensuring enabling and supportive environments.
Delivery of essential services needed by the growing number of older persons.	[Implementation and follow-up].	Implementation and follow-up.

quality of life of the population generally. The Plan conveyed an optimistic view of current extensions of life in good health, the effectiveness of low cost interventions for serious illnesses, and prospects for insurance schemes to finance hospitalizations. There was no specific mention of the health issues arising from poverty, under-nourishment, malnutrition, smoking, malaria, tuberculosis and HIV/AIDS, which impact upon young and old and are the most prevalent and intractable problems for health services and family support in developing countries. Similarly, in relation to housing, transport and private-sector service provision for the aged, the discussion mainly presented a middle-class perspective rather than realistic ways of assisting the impoverished majority in many countries of the region.

Regarding the implementation of policies, the Plan highlighted the need for collaboration of governments, non-government organizations and the private sector together with the importance of encouraging volunteering by and for older persons. It gave special emphasis to the role of non-government organizations:

> The governments are aware of the critical role that non-governmental organizations play. Many of these organizations are pioneers in organizing and delivering services to older persons. In many member countries, they are the core institutions in the national infrastructure for ageing. The governments will encourage the development of a strong non-governmental sector and enhance the participation of these organizations in planning and implementing policies and programmes concerning older persons. Facilitation of the development of this sector should include the formulation of a legal framework for the establishment and registration of such organizations, and the provision of budgetary subvention and assistance in other forms to them. (ibid.: 12).

An ensuing document, *Guidelines on the Implementation of the Macao Plan of Action* (ESCAP 1998b), summarized key points from the Plan through providing concise recommendations on action in relation to the four issues or immediate tasks listed above (see also Table 15.1). The Guidelines further referred to action in specific areas such as dispelling negative stereotypes, supporting the care-giving role of families and improving health, housing, employment and social services – especially home support services (ibid.: 3–10). Overall, the Plan identified wide ramifications of population aging and developed an extensive agenda to assist Asia-Pacific governments in determining their own responses. It also laid the ground for furthering progress by advocating international cooperation in implementing recommendations.

Examples of the proposed ways of achieving this included the exchange of information and expertise by countries and organizations, United Nations' support for country-level initiatives and ESCAP undertaking periodic reviews of the implementation of the Plan (ibid.: 12).

15.2 The Shanghai Implementation Strategy

Following the preparation of the *Madrid International Plan of Action on Ageing 2002*, United Nations regional commissions undertook to formulate regional implementation strategies. These included strategies for Europe (ECE 2002), Latin America and the Caribbean (ECLAC 2003) and Asia and the Pacific (ESCAP 2003). In the same year as the Madrid conference, ESCAP organized a meeting in Shanghai to develop an Asia-Pacific regional implementation strategy for both the Madrid Plan and the Macao Plan. ESCAP's report observed that "there is a great concordance between the Macao and Madrid plans of action" (ibid.: i). Table 15.1 illustrates this, comparing their major stated policy directions as well as that of the *Shanghai Implementation Strategy*. The Madrid and Shanghai documents have similar structures. Although the Macao Plan was organized somewhat differently, the main thrust of the document was population aging and development, including the importance of maintaining health and well-being.

The structure of the documents, and the differences between them, also reflect the absence of agreed policy concepts that might provide a focus for discussion of broad subject areas, such as concepts of aging in place and productive aging. 'Active aging' (see Chap. 12), however, is mentioned in the introduction to the Shanghai Strategy as an encompassing concept: "A national strategy on how to prepare society for the challenges of aging is essential in ensuring that the goals of active aging are achieved." Similarly, one of the key actions in the report states: "Promote policies and programmes that support active aging, which is the process of optimizing opportunities for health, participation and security in order to enhance the quality of life as people age" (ESCAP 2003: 1 and 7).

The *Shanghai Implementation Strategy* presented guidelines on responses to population aging for governments in the Asia-Pacific region. While in accord with the Madrid Plan of Action, it reduced the original 239 recommendations to 66 'key actions'. The report is structured in terms of the topics shown in Table 15.1, providing recommendations for action in each of these areas. Like the other two reports, it made no mention of population policies apart from the need to take population aging into account in policy planning. In relation to population aging and development, the Shanghai Strategy revealed the extensive nature of the task and ways forward, as summarized in Table 15.2.

Taken in isolation, and without advice on accomplishing the recommended actions with minimal resources, many of the key actions seem contingent upon economic development, although this is no panacea if great inequalities persist between social groups. The most practicable means of assisting families and the aged are likely to be low-cost strategies integrated with other aspects of development.

Table 15.2 Aging and development in the Shanghai Implementation Strategy

Objective	Difficulties	Examples of key actions
Mainstreaming aging into development policy.	Lack of funds and staff, lack of staff training, ineffective interdepartmental co-operation.	Increase the efficiency of existing systems, obtain technical assistance, provide training opportunities, promote interdepartmental collaboration.
Promoting integration and participation of older persons.	Many live in rural and remote areas, lack of family and community support, under-recognition of older people's contributions and potential.	Recognize and support the contribution of older people to the family and the community, promote the social and economic participation of older persons.
Provision of social protection and security.	Except for public sector employees, there is little pension coverage, such as for the informal sector and rural areas, or people without employment or earnings. Inadequate tax revenue for universal pension schemes.	Establish sustainable social security systems for the labour force in general and older persons in particular. Gather and utilize data on living conditions, incomes and expenditure to provide a reliable basis for policies on income security.
Alleviation of poverty in old age.	High prevalence of poverty, with older persons among the poorest.	Include older persons as a target group in poverty alleviation; support their capacity to undertake income generation schemes.
Older persons and emergencies.	Older persons, especially those without families, are particularly vulnerable during natural disasters.	Provide special protection to older persons during and after natural disasters and other emergencies.
Promoting positive attitudes towards aging and older persons.	The prevalence of negative stereotypes of older people in the general population and in the mass media – as dependent, frail, troublesome and unable to contribute.	Promote, through media campaigns and school curricula, recognition of the contribution of older persons to society. Encourage the media to promote positive images of aging.
Employment of older persons.	Measures to achieve productive aging, through the continued employment of older persons in the workforce, have yet to succeed.	Provide incentives and remove disincentives for people to remain in the workforce. Seek effective measures to combat unemployment.
Recognizing gender-specific issues in aging.	Women form the majority of people aged 75 and over. They are more likely to be widowed, and to lack income security and skills. They are also the primary caregivers and in many cases work only in the home or in the informal sector.	Enhance support for family caregivers, to allow them to combine work and family life. Promote greater male responsibility in the family. Eliminate gender discrimination and increase the participation of women in the labour force through education and training.

Source: ESCAP (2003: 2–6)

Health care initiatives in developing countries point to some ways forward (see Caldwell 1986; Beaglehole et al. 2008; Kuhn 2010). Given the scope and extent of the range of actions recommended for aging populations in the ESCAP region, the means of implementation are critical to progress or even to keeping pace with population growth and social changes. The Shanghai Strategy emphasizes that governments and other national actors have a major role. The recommended approaches to implementation included data collection and research on the circumstances of older people, encouraging the participation of NGOs, the private sector, older persons' associations and other sectors of civil society, together with regional and international cooperation (ESCAP 2003: 11–12). Despite the importance of these means of implementation, more research is needed on low-cost strategies for improving the circumstances of the aged in developing countries. Successful implementation of plans of action will often depend on such initiatives. Also needed is the identification of priorities for individual countries. Like the Madrid Plan, the Shanghai Strategy identified four major policy directions and recommended actions relevant to each (Table 15.1), but it did not indicate priorities. The absence of this reflects the diversity in levels of socio-economic development within the Asia-Pacific region.

Although priorities necessarily vary between different national settings, as well as between urban and rural regions, three issues stand out as key concerns for the rapidly growing aged populations in Asia's developing countries, namely poverty alleviation, health care, and support for families. All are interwoven with other issues, but in many developing countries progress in these areas could have the greatest benefit for the welfare of older people. This needs to take place in the context of building 'a society for all ages'. As noted in the *Report on the Second World Assembly on Ageing*: "A necessary first step in the successful implementation of the Plan is to mainstream aging and the concerns of older persons into national development frameworks and poverty eradication strategies" (United Nations 2002: 39). For example, in responding to the Madrid Plan of Action, the United Nations Economic Commission for Europe (ECE) made commitments: "To mainstream ageing in all policy fields with the aim of bringing societies and economies into harmony with demographic change to achieve a society for all ages" and "To ensure full integration and participation of older persons in society" (ECE 2002: 1–2).

15.3 Poverty Alleviation

In 2005 about a quarter of the population in less developed regions lived on less than US$1 per day, compared with half of those in the least developed countries (United Nations 2010: 2). The key actions from the Shanghai Strategy that relate most closely to poverty eradication are:

- Provide adequate social protection/social security coverage for the labour force in general, including the agricultural and informal sectors, and older persons in particular, recognizing the role of government as supporter and regulator.

- Include older persons as a target group in poverty alleviation programmes at all levels, including income generation schemes and savings and credit programmes, with emphasis on high-risk groups such as women (ESCAP 2003: 3–4).

Eradicating extreme poverty and hunger is the first of the Millennium Development Goals agreed to in the *Millennium Declaration*, signed in September 2000 by 189 countries (Box 15.1). Poverty commonly entails multiple areas of deprivation including insufficient income, illiteracy and no access to adequate shelter, nutrition and health care, or to clean drinking water and sanitation (Harper 2006: 209–10). Figure 15.1 shows that a high percentage of the developing country populations in Asia have incomes less than $1 per day. Some countries have achieved marked reductions in these percentages since the early 1990s, although having more than $1 per day is no guarantee of deliverance from extreme poverty.

Box 15.1 Millennium Development Goals (MDGs)

The goals and targets of the *Millennium Development Goals* signify a partnership between the developed countries and the developing countries to promote development and eliminate poverty:

> The Goals represent human needs and basic rights that every individual around the world should be able to enjoy – freedom from extreme poverty and hunger; quality education, productive and decent employment, good health and shelter; the right of women to give birth without risking their lives; and a world where environmental sustainability is a priority, and women and men live in equality. (Ban Ki-moon, in United Nations 2010: 3).

Goal 1: Eradicate extreme poverty and hunger.
Target 1.A: Halve, between 1990 and 2015, the proportion of people whose income is less than one dollar a day.
Target 1.B: Achieve full and productive employment and decent work for all, including women and young people.
Target 1.C: Halve, between 1990 and 2015, the proportion of people who suffer from hunger.

Goal 2: Achieve universal primary education.
Target 2.A: Ensure that, by 2015, children everywhere, boys and girls alike, will be able to complete a full course of primary schooling.

Goal 3: Promote gender equality and empower women.
Target 3.A: Eliminate gender disparity in primary and secondary education, preferably by 2005, and in all levels of education no later than 2015.

Goal 4: Reduce child mortality.
Target 4.A: Reduce by two-thirds, between 1990 and 2015, the under-five mortality rate.

(continued)

Box 15.1 (continued)

Goal 5: Improve maternal health.
Target 5.A: Reduce by three quarters, between 1990 and 2015, the maternal mortality ratio.
Target 5.B: Achieve, by 2015, universal access to reproductive health.

Goal 6: Combat HIV/AIDS, malaria and other diseases.
Target 6.A: Have halted by 2015 and begun to reverse the spread of HIV/AIDS.
Target 6.B: Achieve, by 2010, universal access to treatment for HIV/AIDS for all those who need it.
Target 6.C: Have halted by 2015 and begun to reverse the incidence of malaria and other major diseases.

Goal 7: Ensure environmental sustainability.
Target 7.A: Integrate the principles of sustainable development into country policies and programmes and reverse the loss of environmental resources.
Target 7.B: Reduce biodiversity loss, achieving, by 2010, a significant reduction in the rate of loss.
Target 7.C: Halve, by 2015, the proportion of people without sustainable access to safe drinking water and basic sanitation.
Target 7.D: By 2020, to have achieved a significant improvement in the lives of at least 100 million slum dwellers.

Goal 8: Develop a global partnership for development.
Target 8.A: Develop further an open, rule-based, predictable, non-discriminatory trading and financial system. Includes a commitment to good governance, development and poverty reduction – both nationally and internationally.
Target 8.B: Address the special needs of the least developed countries. Includes: tariff and quota free access for the least developed countries' exports; enhanced programme of debt relief for heavily indebted poor countries (HIPC) and cancellation of official bilateral debt; and more generous ODA [official development assistance] for countries committed to poverty reduction.
Target 8.C: Address the special needs of landlocked developing countries and small island developing States (through the Programme of Action for the Sustainable Development of Small Island Developing States and the outcome of the 20-second special session of the General Assembly).
Target 8.D: Deal comprehensively with the debt problems of developing countries through national and international measures in order to make debt sustainable in the long term.
Target 8.E: In co-operation with pharmaceutical companies, provide access to affordable essential drugs in developing countries.
Target 8.F: In cooperation with the private sector, make available the benefits of new technologies, especially information and communications.

Source: United Nations (2011).

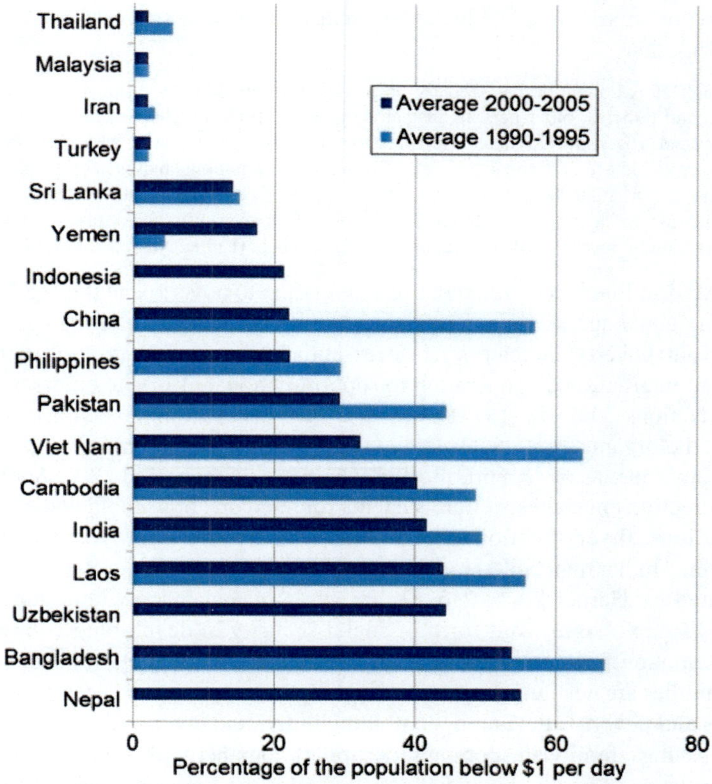

Fig. 15.1 Percentage of the population of Asian countries with incomes less than $1 per day, 1990–1995 and 2000–2005 (Source: United Nations 2011)

The causes of poverty, the means of reducing it, and the feasibility of public and private pension schemes are complex and continually debated (ibid.: 214–5 and 227–30). Nevertheless, it is clear that most developing countries will be unable to afford to construct national social security systems modeled on those of Western countries. In developing countries, coverage of these has mainly been limited to urban groups, such as government employees, while agricultural workers and the urban and rural poor have not benefitted. In developing Asian countries, formal social security coverage ranges from about 9% to 30% of the aged population. Although many countries provide social entitlements, access to those entitlements is unreachable for many of the aged, especially those most in need (ESCAP 2004: 5). Magnus (2009: 165) considered that Asian countries had 10–20 years in which to prepare or remedy their social insurance systems, but concluded that: "no one should imagine this to be either inevitable or successful." As stated in the Madrid Plan:

> Although global attention has recently been focused more actively on poverty eradication targets and policies, older persons in many countries still tend to be excluded from these policies and programmes. Where poverty is endemic, persons who survive a lifetime of poverty often face an old age of deepening poverty. For women, institutional biases in social protection systems, in particular those based on uninterrupted work histories, contribute further to the feminization of poverty. ... Special social protection measures are required to address feminization of poverty, in particular among older women. (United Nations 2002: 18).

The Madrid Plan's recommendations in relation to poverty and the aged included promoting equal access for older persons to employment, developing age and gender relevant poverty indicators as an essential means of identifying needs, and enhancing international cooperation to support national efforts to eradicate poverty (United Nations 2002: 18–19). The Madrid Plan also recommended, as a matter of urgency, the organization of systems to ensure minimum income for older persons with no other means of support, most of whom are women (ibid.: 20). Another suggested direction envisages welfare systems founded on "household and community contributions, diversification of economic activity and improvement of market functioning" rather than built through government programs alone (Burgess and Stern 1991, cited by Harper 2006: 215–6). Transfers of assistance within and between households are a significant form of social security in developing countries, but supplementation from community and government tiers of support is needed when whole families are destitute. Consequently, some envisage a multifaceted approach to pensions and poverty alleviation, combining individual contributions from employment or savings, family and community support, together with government pensions targeting the most needy. Financial assistance for the impoverished aged can have significant flow-on benefits, enhancing intergenerational support and raising the living standards of entire households (ibid.: 216 and 229). Lack of government revenue, however, is but one of the obstacles confronting progress in these directions. Others include expensive and inefficient administrative procedures, as well as non-compliance and corruption – all of which increase developing countries' difficulties in responding to social and economic issues.

For a majority of the aged at any time, later life is not a period of dependency and decline, but one of continuing participation in the family and the community (Heisel 1984: 61). The aged who are not in the mainstream labour force make many contributions to society, or have the potential to do so if afforded opportunities. Communities can benefit from their knowledge, experience and free time through providing opportunities for volunteer work to those who want it, and through recognizing and acknowledging their contributions. Volunteer work, however, will be of little use to people who are struggling for their own existence; their need is for paid work or practical help. In relation to paid employment, another dilemma for less developed countries is whether to maintain the labour force participation of older workers in the modern sectors of the economy – potentially to the disadvantage of younger and better educated people – or to foster older workers' early retirement and prolong their time lived as a possible tax burden.

15.4 Health Care

People in developing countries not only have lower life expectancies than people in developed countries, but also live a higher proportion of their lives in poor health (Lopez et al. 2006: 7). Yet health care for older people in developing countries was long neglected amid much-needed concern for child survival and the health of mothers. The forthcoming increase in dementia cases is part of the more general sustained surge in the numbers of older people with chronic illnesses and disabilities. This is associated with progress towards the third stage of the epidemiologic transition in developing countries. For example, WHO projections indicate that Africa, Asia, and Latin America will have more than 55 million people with senile dementia in 2020 compared with 29 million in 1998 (WHO 1998). Other projections for Asia envisage the prevalence of dementia in China rising from 5.5 million cases in 2005 to 27 million in 2050. The corresponding figures for India are 3.2 million and 16.3 million (Access Economics 2006: v). Senile dementia, due to Alzheimer's disease and other causes, is one of the most incapacitating afflictions of later life and even more moderate cases may require high levels of personal care.

Other chronic illnesses – including heart disease and cancer – are also among the most challenging aspects of population aging. They require provision for a new and expanding group of vulnerable people. They also call for preventive actions to ameliorate trends. The emphasis in developing countries has long been on infectious diseases and primary health care for mothers and infants, which is especially appropriate for countries in the second stage of the epidemiologic transition. Population growth and aging are now bringing a rising need for attention to chronic and degenerative diseases and care of the aged. This trend coexists with the persistent scourge of diseases such as malaria, HIV/AIDS, and tuberculosis. Emerging priorities for health systems are disease prevention and primary care for all age groups. Key actions from the Shanghai Strategy include strengthening primary health care and bolstering family care of dependent relatives:

- Ensure a continuum of health-care programmes based on a primary health systems approach, including locally based health-care practices, health education, health promotion, disease prevention and a coordinated referral system to hospital and other health-care services.
- Improve access to and the quality of long-term care for older persons and develop social support systems to enhance the ability of families to take care of older persons within the family.
- Adopt a comprehensive planning approach taking account of available resources within the community, such as neighbours and volunteers, and direct efforts towards developing interdepartmental as well as intersectoral collaboration. The active involvement of NGOs and the private sector are vital in this regard (ESCAP 2003: 7–8).

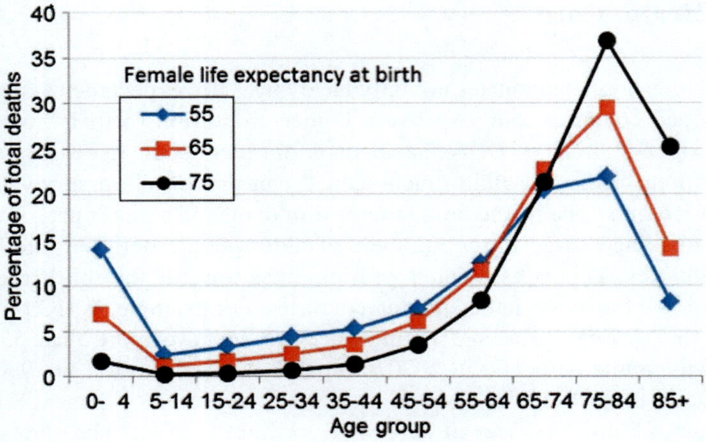

Fig. 15.2 Age distribution of female deaths at different life expectancies (Source: West model life tables (Coale and Demeny 1983; Rowland 2003: 316–319))

The Shanghai Strategy described primary health care as the most efficient and cost-effective way of addressing needs including those of the aged, for whom non-communicable or chronic diseases are the leading causes of morbidity, disability and mortality. Some related key actions entail promoting active aging and healthy lifestyles as well as the development of community based services – which have the potential to relieve the aged care burden on family members (ibid.: 9–10). The Madrid Plan also recommended the strengthening of primary health care services through measures such as training health care workers and social workers in basic gerontology and geriatrics, including traditional medicine in primary health care programs where appropriate, and enabling local communities to provide health support services for older people (United Nations 2002: 27).

Figure 15.2 uses model data on mortality to illustrate aspects of the transformation of the demand for health care in Asian countries. Mortality data are partial indicators of the burden of care because people nearing the end of their lives are especially likely to need health services and personal assistance. Already well under way are the diminishing peak in early childhood mortality and the rising peak in deaths occurring at ages 75 and over. The only constant is that the deaths in the young-old ages (65–74) remain around 20–23% of the total. Over time, differences in national age structures will result in considerable departures from these model figures, which refer only to stationary, zero growth, populations. Nonetheless, as populations experience longer life expectancies the health issues of later life will loom larger and the numbers of deaths at older ages will continue to increase rapidly. The changing age distribution of deaths has major implications for family and community care, because ill health among the elderly contrasts with ill health in early childhood in its causes, treatment and the nature and duration of support required.

15.5 Support for Families

Family care for the aged is a major element of policy proposals for health care in developing countries, and indeed is one of the foundations of overall responses to aging. The Shanghai Strategy discussed the question of supporting the family in its caring responsibilities mainly in the section on "Ensuring enabling and supportive environments" (ESCAP 2003: 8–11). In the past, emphasizing family responsibility provided justification for inaction, but enlightened approaches now recognize the importance of supporting the family, especially in relation to aged care and poverty alleviation. Although the Shanghai Strategy describes the role of governments as vital in improving health services, some developing country governments in the region have had only a minor role in health care for the aged. The family has borne much of the burden, with religious, community and other organizations providing a little additional support. The Shanghai Strategy described community resources for the long-term care of dependent older persons as "scarce" and constituting "a serious challenge".

The importance of supporting the family is reflected in the Strategy's relatively large number of key actions relevant to this issue, one of which was quoted in the previous section:

- Provide integrated care services which allow individuals to remain in their communities for as long as possible.
- Develop and/or strengthen a range of community-based services that support older persons with or without families and family caregivers in which caring responsibilities can be shared among individuals, families, communities, NGOs and government.
- Promote and provide direct support to family caregivers in the form of material aid, tax reduction, subsidized housing or training on home care and develop an integrative model combining both formal and informal care in enhanced community care to help to relieve the care burden of family members.
- Promote and encourage community-based programmes which assist and act as relief mechanisms for family members and caregivers.
- Promote support systems for elderly caregivers of people living with HIV/AIDS in general, AIDS orphans and older persons living with HIV/AIDS (ESCAP 2003: 8–10).

The Madrid Plan made similar recommendations (United Nations 2002: 32ff), but the emphasis on the family is stronger in the Shanghai Strategy. Importantly, the latter described prospects as unfavourable because caregiver support was a relatively low priority in parts of the Asia-Pacific region. The Shanghai Strategy also noted that: "the numbers of older persons at high risk of dependency and disability are increasing at the same time as the ability of families to provide care is decreasing". Moreover, the HIV/AIDS pandemic is adding greatly to the burden on caregivers, including elderly caregivers (ESCAP 2003: 7–10). Given the diversity of family living arrangements and traditions in Asia, as well as the extent of demographic, social and economic changes, it is misleading to assume that the family has the capacity to function independently as the provider of aged care. Participants at a 2004 seminar on the Shanghai Strategy concluded that "there is a need to build on existing informal support systems with formal schemes" (ESCAP 2004: 5).

Finally, another concern, raised in the Madrid Plan of Action 2002, included whether backing for aging in place sometimes arose from financial considerations and the assumption that families would provide the bulk of the care:

> Without adequate assistance, family caregivers can be overburdened. In addition, formal community care systems, even where they exist, often lack sufficient capacity because they are poorly resourced and coordinated. As a result, residential care may be the preferred option of either the frail older person or the caregiver. In view of this range of issues, a continuum of affordable care options, from family to institutional, is desirable. Ultimately, the participation of older persons in assessing their own needs and monitoring service delivery is crucial to the choice of the most effective option. (United Nations 2002: 35).

Women are likely to experience to a greater extent than men the consequences of disadvantaged families and deficiencies of aging in place. This reflects that women have greater longevity and higher risks of chronic illnesses, discrimination and marginalization. They are also more likely to be caregivers. Significant for the future of aging and the family in Asia is an emerging recognition of over-reliance on informal and family sources for long-term care (ESCAP 2003: 8–9). This acknowledges changes within the family as well as its growing and already greater burden of responsibility for aged care than ever existed in the past. Better information is needed about the nature and extent of intergenerational care, its adequacy, the difficulties faced and the availability of formal and informal assistance, including when family support is unavailable. There has been increased investment in in-depth and longitudinal surveys to answer such questions (United Nations 2005: 111).

15.6 Conclusion

This chapter has sought to discern, from the many policy recommendations in the Shanghai Strategy, three broad issues most relevant to developing countries generally. Through a focus on matters directly relating to better survival, the 'short list' omits other issues often deemed to be of prime importance, such as ageism and gender inequality. It also omits positive aspects of the growing numbers of healthy, independent older people, such as their present and future potential contributions society. Although short lists inevitably vary, it is clear that the wide-ranging nature of international policy documents on aging calls for attention to prioritizing recommendations for particular countries. The United Nations Department of Economic and Social Affairs engaged in a more extensive exercise of this type in its *Guide to the National Implementation of the Madrid International Plan of Action on Ageing* noting that: "The scope of the Madrid Plan is rather broad, covering a variety of topics and incorporating 239 separate recommendations; this Guide addresses the most crucial areas requiring particular policy attention" (United Nations 2008: 1).

The Guide nonetheless traversed a broad agenda including social and economic development, social protection, health policy, long-term care and social inclusion. As in the Madrid Plan, a key theme of the Guide was the integration of aging and the aged into mainstream policy making. An important aspect of this is enhancement of the status of the aged in Asian societies. The Guide emphasized that "the most fundamental requirement for change involves eliminating negative attitudes and stereotypes associated with older persons" (ibid.: 7). Perceptions of low status contribute to ageism, expectations of disengagement of the aged from society, under-resourcing of their needs and failure to engage in proactive planning for their future.

References

Access Economics. (2006). *Dementia in the Asia Pacific Region: The epidemic is here.* Report for Asia Pacific Members of Alzheimer's Disease International. Canberra: Access Economics. http://203.89.197.244/common/files/NAT/20060921_Nat_AE_FullDemAsiaPacReg.pdf. Accessed May 2011.

Beaglehole, R., et al. (2008). Alma-Ata: Rebirth and revision 3 – Improving the prevention and management of chronic disease in low-income and middle-income countries: A priority for primary health care. *The Lancet, 372*, 940–949.

Burgess, R., & Stern, N. (1991). Social security in developing countries: What, why, who and how. In A. E. Dreze, J. Hills, & A. Sen (Eds.), *Social security in developing countries* (pp. 41–80). Oxford: Clarendon Press.

Caldwell, J. C. (1986). Routes to low mortality in poor countries. *Population and Development Review, 12*(2), 171–220.

Coale, A. J., & Demeny, P. (1983). *Regional model life tables and stable populations.* New York: Academic Press.

ECE (Economic Commission for Europe). (2002). *Regional implementation strategy for the Madrid International Plan of Action on Ageing 2002*. UNECE Ministerial Conference on Ageing, Berlin. http://www.un.org/ageing/impl_map.html. Accessed Aug 2010.

ECLAC (Economic Commission for Latin America and the Caribbean). (2003). *Regional strategy for the implementation in Latin America and the Caribbean of the Madrid International Plan of Action on Ageing*, Santiago. http://www.un.org/ageing/impl_map.html. Accessed Aug 2010.

ESCAP (Economic and Social Commission for Asia and the Pacific). (1998a). *Plan of action on ageing for Asia and the Pacific*. Bangkok: ESCAP. http://www.unescap.org/ageing/macau/plan_of_action.htm. Accessed Dec 2009.

ESCAP. (1998b). *Guidelines on the implementation of the Macao Plan of Action on Ageing for Asia and the Pacific*. Bangkok: ESCAP. http://www.unescap.org/ageing/macau/guidelines.htm. Accessed Dec 2009.

ESCAP. (2003). *Shanghai Implementation Strategy: Regional implementation strategy for the Madrid International Plan of Action on Ageing 2002 and the Macao Plan of Action on Ageing for Asia and the Pacific 1999*. Bangkok: ESCAP, E/ESCAP/1280.

ESCAP. (2004). *Regional review: Report of the regional seminar on follow-up to the Shanghai Implementation Strategy for the Madrid and Macao Plans of Action on Ageing*. Bangkok: ESCAP. http://www.unescap.org/esid/psis/meetings/AgeingJuly2006/Macao%20report%2026%20Jan%20draft.pdf. Accessed Jan 2010.

Harper, S. (2006). *Ageing societies: Myths, challenges and opportunities*. London: Arnold.

Heisel, M. A. (1984). Aging in the context of population policies in developing countries. *Population Bulletin of the United Nations*, (17), 49–63.

Kuhn, R. (2010). Routes to low mortality in poor countries revisited. *Population and Development Review, 36*, 655–692.

Lopez, A. D., Mathers, C. D., Ezzati, M., Jamison, D. T., & Murray, C. J. L. (2006). *Global burden of disease and risk factors*. Washington, DC: Oxford University Press and the World Bank.

Magnus, G. (2009). *The age of aging: How demographics are changing the global economy and our world*. Singapore: Wiley (Asia).

Rowland, D. T. (2003). *Demographic methods and concepts*. Oxford: Oxford University Press.

United Nations. (2002, April 8–12). *Report of the second world assembly on ageing*, Madrid. New York: United Nations.

United Nations. (2005). *Living arrangements of older persons around the world*. United Nations Department of Economic and Social Affairs, Population Division. New York: United Nations, ST/ESA/SER.A/240.

United Nations. (2008). *Guide to the national implementation of the Madrid International Plan of Action on Ageing,* New York: United Nations. http://www.un.org/esa/socdev/ageing/documents/papers/guide.pdf. Accessed Jan 2010.

United Nations. (2010). *The millenium development goals report 2010*. New York: United Nations, Department of Economic and Social Affairs. http://www.un.org/millenniumgoals/pdf/MDG%20Report%202010%20En%20r15%20-low%20res%2020100615%20-.pdf. Accessed Jan 2011.

United Nations. (2011). *Millenium development goals indicators: The official United Nations site for the MDG indicators*. United Nations, Department of Economic and Social Affairs. http://mdgs.un.org/unsd/mdg/Host.aspx?Content=Indicators/OfficialList.htm. Accessed May 2011.

World Health Organisation. (1998). Population ageing: A public health challenge. *Fact Sheet* No. 135. https://apps.who.int/inf-fs/en/fact135.html. Accessed June 2011.

Chapter 16
Prospects

... the decline of the labour force poses a far greater threat than ageing alone.

(Heigl 2004: 287)

16.1 Trajectories

Almost all developed countries have entered their demographic autumn, when youthful age structures are finally past. This situation is unprecedented – its problems are often unrecognized or unresolved and its potential is unrealized. Yet the world as a whole is still in the take-off stage of population aging as only in the second half of this century will the process culminate in many developing countries. In developed countries population aging will peak much sooner – during the second quarter of the century. In some, the brief autumn so far experienced could quickly give way to a demographic winter of long-term population decline and hyper-aging. Achieving and stabilizing national age structures with favourable characteristics is a priority for the first half of the twenty-first century. Even if demographic goals are only partially attainable, early interventions could do much to alleviate national situations. Unexpected developments, including economic crises, distract attention from responding to demographic and social changes, as do other long-standing concerns for many societies – especially global warming, globalization, unemployment, national debt and poverty. This chapter focuses on prospects for aging societies in relation to theories relevant to the long-term future and initiatives conducive to demographic and fiscal sustainability. The latter includes Third Sector development, which is likely to have an expanded role in securing individual and family welfare in vulnerable situations. The next, concluding, chapter addresses prospects with reference to particular national situations, comparing countries with the oldest and largest aged populations in terms of their exposure to risks from aging, their resilience in the face of changes, and factors that make the greatest positive difference to societies' experiences of population aging.

D.T. Rowland, *Population Aging: The Transformation of Societies*,
International Perspectives on Aging 3, DOI 10.1007/978-94-007-4050-1_16,

16.2 Theories of Long-Term Change

Theories of long-term change, especially transition theories, provide a point of departure for interpreting trends and assessing future directions. This book has applied such theories, especially in the chapters on demographic changes, health and the family. Yet there are other more general, global theories that provide insights into the possible long-term outcome of population trends. Two such theories are concerned with homeostasis and convergence. A shift to a demographic winter was rarely contemplated in the past, so influential was the implicit belief that populations which ceased to grow would at least continue to perpetuate themselves. Demographic transition theory consolidated this belief, but low birth rates since the 1970s, together with contemporary theories of fertility change, have contradicted it. Nevertheless, a related demographic theory of long term change, namely homeostasis theory, still suggests an ultimate end point where national populations have close to constant numbers and age structures. Such an eventuality is often posited as desirable for the world and for individual countries.

Homeostasis envisages that social, cultural and economic institutions create a self-regulating system which tends to maintain a population growth rate around zero in the long run; otherwise the population would outstrip its resources (Wilson 2003: 493; Wilson and Airey 1999). At least partly in accord with this are population projections which assume that global population growth will cease before the end of the twenty-first century. Evidence of homeostasis, however, derives mainly from the demographic transition's pre-transition stage which encompasses all but the last 200 years or so of human history. Low rates of population growth, averaging little more than zero, are thought to have been typical before the transition began in Europe during the second half of the eighteenth century, in conjunction with the industrial revolution. The human population is still far from the end of the current main phase of population growth, because of continuing high birth rates in many developing countries together with the procession of larger generations into older ages in all countries. The latter phenomenon, which produces population momentum, is responsible not only for much of the global population's relentless growth but also for the unstoppable impact of population aging. Thus, even if there is an inherent protracted tendency towards homeostasis it will not modify the growth in the numbers and percentages of the aged during coming decades. Yet if policies are supportive of near replacement level fertility and moderate net immigration, around mid-century some countries with currently low population momentum could approach a situation akin to homeostasis.

Such an outlook is compatible with convergence theory, which revises ideas from classical transition theory to recognize forces of change since the late twentieth century. The theory envisages that societies are converging in their demographic characteristics because of globalization and common goals in relation to national development. Jones (1993: 1 and 5) argued that:

> Globalizing processes of industrialization and Westernization are creating a retreat from diversity in human experience. …On a world scale, cessation of population growth is not only desirable but inevitable. The only issues are how long it will take to happen and how it will come about.

Particular facilitating factors that Jones cited include: (i) modern transport and communications; (ii) the homogenization of the content of school text books, television programs and movies; (iii) elites of developing countries becoming similar to those of the West in terms of education, lifestyles and consumption patterns; (iv) the ideal of 'developmentalism' – the aim of bringing socio-economic development to all parts of the world; and (v) the formation of large economic communities, such as the Association of Southeast Asian Nations (ASEAN), which comprise a further force for mixing and homogenization.

Thus, according to convergence theory the economic rationality and competitiveness of the industrial mode of production, in a globalized economy, cause all industrial populations to experience a similar pattern of development. This is characterised by strong conformist pressures on individuals, families, and national economic and social policies. If demographic characteristics follow the convergence of economic characteristics and social institutions, uniform demographic patterns should develop, with any contrasts attributable simply to different stages of development (Coleman 1998: 7). Yet any ultimate convergence to zero growth would entail greater diversity in the representation of age groups. Whereas triangular age structures have a predominance of children and young adults, rectangular age structures associated with low and balanced birth and death rates have a more even representation of all age groups. For the first time in history a society of all ages would emerge, one requiring to be also "a society for all ages." Nevertheless, ethnic diversity and intermixing will be further characteristics of supposedly converged societies that have had a long history of immigration and of associated linguistic, religious and cultural heterogeneity.

Considering zero growth as a convergence point raises questions about the population size at which stability should be achieved. The founders of the One Child Policy believed that China's future population had to be smaller than the existing one in order to avoid widespread starvation. Some Western writers also favour a low population target for their own countries in order to limit pressure on resources and the environment. Yet others favour a high target, or even none at all, on the grounds that population growth stimulates economic growth and resource creation. Past debate about optimum population revealed many criteria for defining a future target population, including demographic and environmental ones, as well as food supplies and living standards. Although conservationists now commonly advocate smaller numbers as desirable for most countries, population decline has many adverse repercussions for the age structure, the labour force, the economy and society generally. Future debate will need to examine population goals in light of new prospects for self-reinforcing population decline and unexpectedly high levels of aging. Ultimately, stabilizing population size and age structure will represent the middle ground in addressing a range of concerns. A zero growth population would have more options in determining its future than a declining population. The latter would be severely disadvantaged in providing for its own citizens because of rising dependency and a shrinking labour force.

In relation to population aging, the most influential aspect of convergence has been the trend towards the two child family becoming the most commonly perceived norm, but significant differences in birth rates persist even between developed countries where convergence might be expected to be most pronounced. Although surveys generally confirm the ideal of two-children per couple, stated preferences are often unrealized and birth rates at different levels below replacement have varying consequences for the degree of aging and the shape of the age profile. While convergence theory implies that common approaches to social issues are becoming more feasible, continuing demographic and social diversity call for varied responses. Convergence theory's focus on similarities distracts attention from differences that greatly affect countries' prospects. For example, even though there has been convergence towards older age structures important differences persist, such as apparent in regional differences within Europe – between east and west and north and south (see Chaps. 3 and 17). Other evidence of non-convergence includes the unevenness of falls in rates of fertility and mortality in different places (Dorius 2008), the persistence of poverty and socio-economic inequality (Wilson 2001: 168), and the resurgent influence of religion in some societies.

16.3 Stabilization

At present, trends in national age structures have no single convergence point but, around the middle of the century, convergence towards the model age profile introduced in Chap. 3 is a possibility for countries that currently have low population momentum (Fig. 16.1). Convergence to the model, stabilized age structure can occur only after population aging has largely run its course. In classical demographic transition theory the rectangular age structure is the expected end point of age structure evolution. English-speaking countries and countries in Western and Northern Europe are relatively well-positioned to achieve nearly stabilized rectangular age structures similar to the model age structure (see Chap. 17). Progress towards this end point is, demographically, the least disruptive form of population aging, although it still requires substantial adjustments in societies through time.

There is growing interest in population stabilization as a long term policy goal. This goal is associated with achieving sustainable development and implementing circumspect approaches to climate change and conservation of resources and the environment. In demographic terms, a stable population is one that is growing or declining at any constant rate, which includes a rate of zero. In conjunction with low birth and death rates, zero growth is the only rate that will maintain a rectangular age structure. Constant rates of growth or decline will produce age structures which, in the long run, will have wide or narrow bases respectively. Net immigration can modify these basic forms, depending on the volume and duration of flows.

Although the notion of stabilization can include a range of growth rates and types of age profiles, the zero growth model is in the middle of the range of alternatives and has several advantages. It entails stabilization of population numbers, which is consistent with the notion that neither global nor national population growth is sustainable indefinitely. It has a constant age structure which avoids the demographic shock waves associated with cohorts of contrasting size reaching different

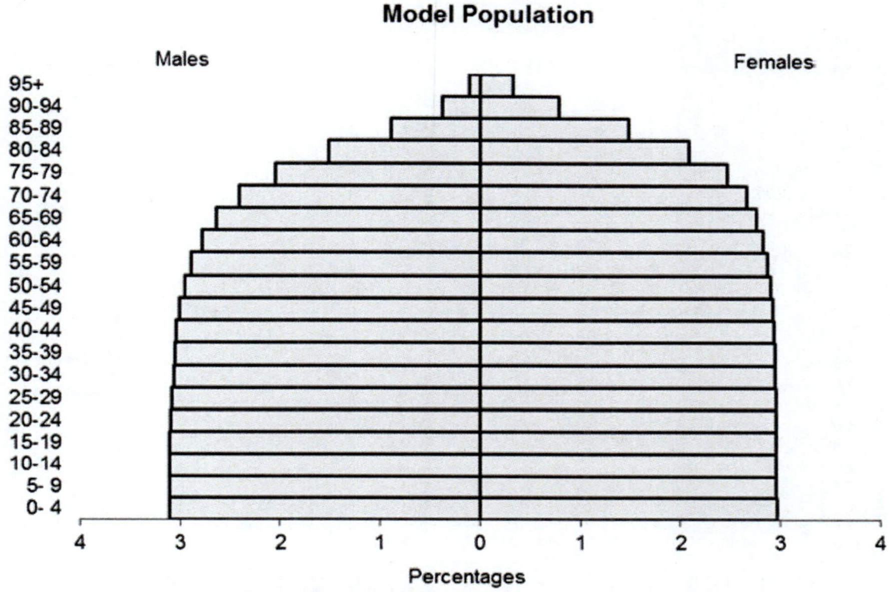

Ages	%
0-14	18.2
15-64	59.3
65+	22.5
Total	100.0

Demographic Indices	
Total fertility rate	2.07
Percentage aged 65 and over	22.5
Total dependency ratio (children and aged per hundred 15-64)	68.7
Aged dependency ratio (aged per hundred 15-64)	38.0
Ageing index (aged per hundred children)	123.8
Life expectancy at birth, both sexes	82.1
Female population momentum	0.0

Fig. 16.1 The model population

stages of life. Also, the percentage of the population in the working ages remains close to 60, much higher than in populations where aging is unrestrained. No national population will ever match the model exactly, because it would require demographic rates to remain constant through time, but approximations are possible. It is a relevant consideration in long-term policy along with ecological sustainability, fiscal sustainability and considerations of population distribution and size. The model also provides a benchmark with which to compare current and projected age structures. Population aging, however, poses considerable challenges for all societies, including those converging towards the characteristics of the model population. Even in the model population the aged outnumber children and dependency ratios are high.

For most societies, zero growth is neither attainable nor desirable immediately because high growth in the older ages is being accompanied by slow growth or decline in the working ages (Fig. 16.2). While the numbers in the older age groups continue to expand, economic sustainability will depend on maintaining labour force size, productivity and participation. In OECD countries the mainstream

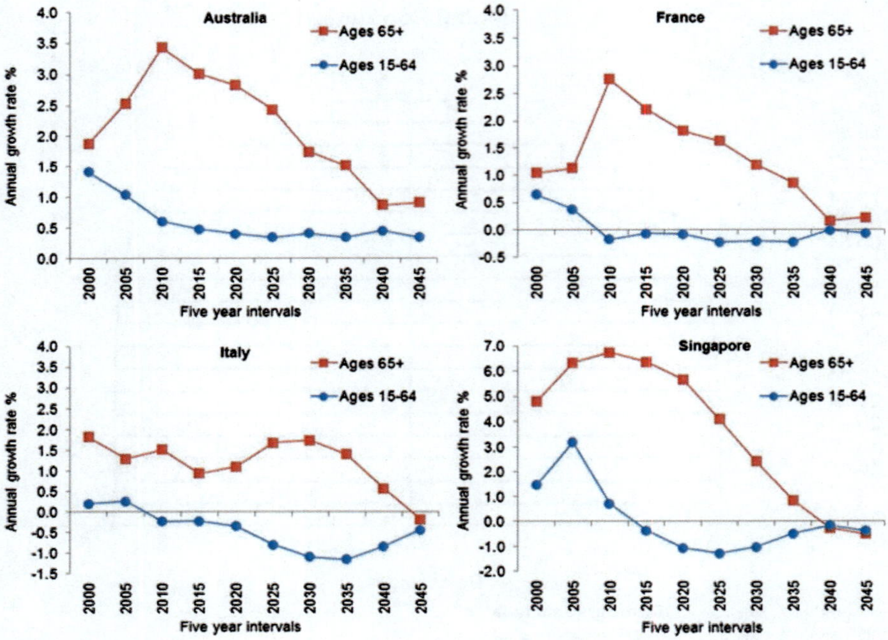

Fig. 16.2 Annual population growth rates at ages 15–64 and 65+, selected countries 2000–2005 to 2045–2050 (Source: Calculated from United Nations 2009a)

view has been that, over coming decades, continuing economic growth will be necessary to meet the needs of aging populations (OECD 1998: 9 and 15). This poses a dilemma amid concerns about climate change, resource depletion and environmental degradation. Although climate change makes the curbing of population growth seem a necessary response, a demographic approach to addressing climate change would probably be less effective in the next 25 years than measures to reduce carbon emissions. Prospects for achieving prosperity without labour force growth should improve as aging runs its course although, if life expectancy continues to increase, gradual population aging will persist long after the main phase is over.

Thus, despite its advantages in the long term, a stationary, stabilized age structure is not possible in the short term and it is not a 'solution' to current concerns about meeting the aspirations of a society while protecting the environment and building the economy. Notions of convergence and stabilization should not distract attention from the important point that countries have yet to experience the principal decades of population aging and associated social and economic changes. The Madrid Plan of Action on Ageing established a wide-ranging agenda for reform but said little about population policies, economic policies and approaches to financing the costs of population aging. Ensuing sections discusses responses in these areas that can contribute to making population aging more sustainable.

16.4 Demographic Sustainability

The main obstacle to stabilizing the age structures of developed countries, and the most problematic demographic change affecting developed countries in the twenty-first century, is negative population momentum. This produces an impetus towards rapid population decline and extreme aging. As discussed in Chap. 2, momentum measures how much population size would change in a country between embarking on a trajectory towards zero population growth and achieving it. If negative momentum has been prolonged, decades of renewed family building would be needed to return to zero growth. Also, the greater the duration of negative momentum, the smaller the ultimate size of the population when and if zero growth is attained.

Although fertility close to replacement level is frequently seen as a prerequisite for demographic and economic sustainability, in the medium term a shift from sub-replacement to near-replacement fertility rates would tend to have a negative effect on economic growth. This is because it would increase the ratio of dependants to workers and reduce for a time the labour force participation rates of mothers (Costello 2002). Thus in European countries with very low fertility, a return to higher birth rates would not have immediate economic benefits in terms of labour force participation. It could take 20 years or more for a higher birth rate to have a direct impact on the labour supply, although subsequently there would be substantial gains.

In the meantime, international migration would be the only component of population change that could contribute to the labour supply. International migration to address labour force deficits is of considerable importance in a number of countries. From 2010 to 2050, the main destinations for international migrants are expected to be the United States, Canada, the United Kingdom, Spain, Italy, Germany, Australia, and France. Over the same period the main countries of origin are likely to be Mexico, China, India, the Philippines, Pakistan, Indonesia and Bangladesh (United Nations 2010: 32).

Japan illustrates the most extreme impacts of negative momentum. United Nations estimates and medium variant projections imply that Japan's population already had negative momentum of −21% in 2010, potentially expanding to −30% by the 2030s. Between 2000 and 2050 Japan's population is projected to fall from 127 million to 102 million. The total of 102 million would be impossible to maintain, however, because the majority of Japan's population in 2050 would be beyond the normal reproductive ages and only a minority of its population in 2050 would contribute to future births. The longer Japan delays any return to replacement level fertility, the smaller the population size it can attain in the long run. The momentum of decline is cumulating in Japan's age structure and could have repercussions for many decades ahead. Longer range projections from Japan's National Institute for Population and Social Security Research imply an outlook of irreversible population decline in Japan with projections ranging between a 'medium' figure of 48 million in 2100 and a 'high' figure of '64 million' (Kaneko 2008: 50 and 53).

Japan's total fertility rate in 2010 was 1.4, a figure attributed both to higher proportions of young women remaining single, and to declining marital fertility.

This has occurred in the context of higher labour force participation of women while their family roles have remained traditional. Postponement of marriage in Japan has not been compensated by an increase in cohabitation (United Nations 2004: 4). Prospective population decline may prompt acceptance of more foreign immigrants, but there is little support for immigration at levels high enough to alter the current trend in the age structure (Roberts 2008: 770). Indeed with such a low birth rate, the volume of migration needed to offset the birth deficit would be unattainable even in a country more disposed to immigration.

Greater attention is being given to encouraging higher birth rates. Some scope for this is perhaps evident in survey data suggesting that the ideal number of children per family in Japan has remained fairly constant at an average of 2.5 since the 1970s, but circumstances affecting marriage and family formation have, since the early 1990s, resulted in TFRs of less than 1.5 (Ogawa 2008: 822). Contemporary low fertility in Japan is associated with changes in attitudes and women's roles similar to those observed in Western countries, together with gender inequality and insufficient child care facilities (Olschleger 2008: 37; Coulmas 2008). The agenda for pronatalist family welfare reform is formidable because of many obstacles: lack of support in the home and the workplace for working mothers; income and career sacrifices ensuing from motherhood; stresses and costs of educating children; the still substantial expectations of family care for the disabled and frail; the high cost of housing; and uncertainties in the labour market. Moreover, government expenditure on children and families accounts for less than 4% of the social security budget, compared with 70% for the elderly. Child allowances remain well below Western European levels (Coleman 2008: 749 and 759–60).

Despite this, family policies in Japan are now conceived as "measures to counter the declining birth rate." Japan's Angel Plan (1995–1999) aimed to provide child-care support for working mothers as well as rectifying future labour shortages through higher participation of women. Encouragement of fathers' involvement in childrearing was also part of the Angel Plan. Its successor, the New Angel Plan (2000–2004) had some similar aims but further emphasized family-friendly workplace changes, such as shorter working hours and child-care leave. It also aimed to reduce the gendered division of labour and the prioritizing of work over family life – which were thought to be causes of the low birth rate. Subsequent policy measures have sought to reinforce family friendly measures and mobilize company support for childrearing, although problems have included a "tepid" response from employers and inadequate funding of policy goals (ibid.: 745–5 and 751–3). Cortazzi (2002: 12) argued that in Japan "Much more needs to be done to create a society in which women want to have children." An initiative towards this goal is Japan's *Basic Law for a Low Birth Society* (2004) which states that the declining birth rate is a problem "unprecedented since the dawn of history." According to this Basic Law there are obligations on national governments and other sections of society to "support a family-friendly work environment, and create a society where people dream of family life and can bear and raise children with peace of mind" (Coleman 2008: 757). Although the "dream" of having children has been criticised as incongruent with individual choices about marriage and childbirth, current wide-ranging pronatalist measures include teaching young people the significance of children and families, and lessening structural obstacles to family formation – such as lack of stable employment and the breakdown of traditional community support networks. With the birth rate remaining perilously low, Japanese family policies have had little pronatalist impact, although they at least represent some positive action amid concerns about the breakdown of the family (ibid.: 761). Other policy responses to aging in Japan have given attention to fostering active aging and the participation of the aged in society, as well as more sustainable health and social security provision.

Population decline is of most concern when it is rapid and occurs in conjunction with rapid population aging, as in Japan. Yet even moderate decline may be viewed negatively because, other things being equal, a smaller population may decrease domestic demand, reduce the taxation base for building and maintaining infrastructure and decrease productivity through lesser economies of scale and division of labour (Coleman and Rowthorn 2011: 226–227). Contradictory evidence is that Germany's GDP has continued to grow despite falling population and, on a global scale, there is no evidence of a positive relationship between population size and GDP per capita. Although total GDP tends to expand with population size, this has no necessary bearing upon individual welfare: the population size sought by strategists and governments may be quite different from the population size that optimizes individual welfare (ibid.: 230). Population stabilization or even slow decline are increasingly perceived as desirable in a time of climate change and other mounting environmental pressures. Amid such concerns, policies that raise labour force participation and productivity can be a more efficient way of addressing labour shortages than increasing population size (McDonald and Kippen 2001: 19). The OECD's 2005 *High-Level Forum on Ageing and Employment Policies* emphasised the need for longer working lives, especially greater labour force participation at ages 50–64. The Forum saw this as the key issue in European countries (Tobback: 2005). Opinion is divided, however, on the question of the productivity of an older labour force (MacKellar 2003: 95; Prskawetz 2005; Skirbekk 2005; Lindh 2005).

Government-sponsored increases in annual births have been attempted through paying one-off financial incentives for childbearing. Although this strategy may have positive effects on the birth rate in conjunction with other family friendly measures, it is insufficient on its own. Financial incentives tend to have a greater influence on the timing of births – as people seek to avail themselves of the handout – than on average completed family size. This is because the value of such payments is usually miniscule compared with the long-run costs of parenting, especially the loss of earnings due to withdrawal from the labour force to care for children. 'Baby bonuses' at least provide public endorsement of the value of childbearing, but genuinely restoring social rewards for parenthood can occur only through long-term amelioration of the financial penalties of parenthood, such as through subsidised childcare, tax concessions, and flexible working hours. Other measures supportive of parenthood include reducing unemployment among young adults, providing assistance with housing costs, and removing obstacles to the employment and career paths of married women and mothers. The public costs of aging will increase unless couples can achieve their desired family building goals, and families can maintain their ability to provide care for dependents. According to Eurobarometer public opinion polls, in almost all European countries the most popular choices are to have two or three children (United Nations 2004: 6).

16.5 Fiscal Sustainability

Issues concerning fiscal sustainability and intergenerational equity are gaining increasing recognition in national policy making. Most OECD countries now produce long-term fiscal projections for intergenerational reporting, although only a few embarked on such exercises before 1999 (Swan 2010: 86–87). They are essential for forward planning. The United Nations, in its *Human Development Report,* as well as the European Commission and other organizations, publish annual or biennial reports on economic, social and environmental sustainability (see Swan 2010: 88 Table 6.2). In anticipation of demographic pressures on government spending, fiscal consolidation – a policy aimed at reducing government deficits and debt accumulation – has been an objective in most OECD countries since at least the early 1990s (OECD 2000: 33). This objective has been largely unfulfilled as crises relating to debt burdens in the Europe and the United States have demonstrated. In the aftermath of the Global Financial Crisis, the OECD (2010) re-emphasized the importance for countries of adopting "a fiscal recovery plan to reduce debt burdens before long-term pressures (especially aging and health care) hit with full force" (OECD 2010: 18). Although child dependency ratios are declining in many countries, this reduction is insufficient to offset the rising costs of population aging. Generally, the costs of supporting older persons are higher than those of supporting children and adolescents (United Nations 2009b: 2).

World-wide, only about a quarter of the labour force is currently accruing pension rights, and four out of five older persons do not have any pension coverage (World Bank, cited by United Nations 2009b: xii). Moreover, the funding of pension schemes in some countries is becoming unsustainable. An emerging realization has been the need for more diversified systems of pension provision:

> Many reforms have been introduced that affect pensions, strengthening of work incentives, employability of older workers, the health and care needs of an older population, and the operation of financial markets in dealing with large increases in private pension savings. A critical challenge has been to slow, and eventually reverse, trends towards a shrinking portion of life being spent in employment and ever-longer periods spent in retirement. Fiscal consolidation has also been a major driver of reform and a more diversified system of retirement income will be the main outcome – more balance among private and public pensions, taxation and, especially, earnings. (OECD 2000: 3).

The benefits of diversification are the spreading of risks, giving individuals more choices, and attaining greater burden-sharing between generations. The mix of arrangements varies greatly from country to country, although poverty alleviation is becoming the main purpose of public pensions, with consequent reductions in more generous approaches (OECD 2000: 48ff). Increases in expenditure on pensions will not necessarily mean increases in the cost measured as a percentage of GDP, which depends on the rate of economic growth. Cross-national differences in expenditure on pensions are more a function of differences in coverage and generosity than of aging. Pension generosity has been greatest in Southern Europe, where benefits are related to previous incomes, and lowest in English-speaking countries where means-testing and poverty alleviation are central. Countries with generous provision have the lowest labour market activity levels and some have very high government debt (Castles 2004: 126–138).

The most thorough-going innovations have entailed movement away from the principle of universal age-based entitlement to a government age pension funded from taxation, and its replacement with needs-based entitlement, together with greater emphasis on life-time saving by individuals to finance their own retirement incomes. Successful adjustment to population aging in terms of income provision will require a shift from the present situation in high income countries, where most retired people rely on a government pension as their main source of income, to a new situation where only a needy minority receive a government pension. The majority would have pensions funded from their own savings through government or employer sponsored schemes, or through private arrangements. Decades of saving are needed to accumulate adequate pension capital and the long lead times will delay increases in the proportions of older people with self-funded retirement incomes. It is unlikely, however, that most people will be able to accumulate sufficient savings to achieve a retirement income greater than the age pension, because paying for housing and providing for a family necessarily take precedence during working life. Women experience particular disadvantages in saving for retirement on account of interrupted workforce participation. Divorces and remarriages also extend people's financial obligations. The financial risks associated with private pension schemes will make industry regulation and compliance of mounting importance as self-provision becomes more prevalent. Unfortunately, some observers anticipate growing retirement income insecurity for workers (Pozzebon 2004). For example, Britain has long had self-funded retirement through employer based pension schemes, but people have sometimes received only a fraction of the predicted pension. The Global Financial Crisis, together with ongoing high levels of government debt, brought about a dramatic world-wide decline in the value of retirement investments.

Preparing the way for more prevalent needs-based pensions and self-provision have been the abolition of mandatory retirement at a designated age, increases in the age of eligibility for a government-funded age pension, the equalizing of the pensionable ages for men and women, and the application of means-testing. Eligibility for a pension is usually conditional upon reaching a particular age after a specified period of contributions, such as 30–40 years of employment. In 2009, men were eligible for a full pension at age 65 or over in about 60% of developed countries. The earliest pensionable ages, of less than 60 years, mainly occurred in developing countries because of lower life expectancies. The trend towards the same pensionable ages for men and women was also most pronounced in developing countries (ibid.: 40–42).

Raising the age of eligibility for a pension has been controversial in some European countries where, with the support of trade unions, tax-payers who contributed to the pensions of older generations have stoutly resisted being deprived of equivalent entitlements. France is relatively well placed demographically, yet it has introduced measures to increase the retirement age from 60 to 62 in 2018, to meet future economic concerns arising from population aging, despite the measures' unpopularity reflected in widespread public opposition. In the United States, the United Kingdom, Australia and New Zealand there was less opposition to earlier initiatives to increase the pension age, whether because of long lead times, incremental changes, extensive consultation and financial incentives for later retirement. Concerns about rising numbers of people retiring early with disability pensions has also led to greater emphasis on assessment of functional capacity, rehabilitation and assistance with retraining and re-entering the labour force (OECD 2000: 19–23). Higher retirement ages can lead to inequities and hardship for some, such as manual workers for whom the physical demands of their employment become less manageable as they get older. Also, they may have worked and paid taxes for more years than people who spent longer in full-time education.

In addressing sustainability issues, further strategies likely to become increasingly important through time are removing incentives for early retirement, education to counter age-discrimination by employers and young employees, and expanding opportunities for older workers through retraining and offering part-time or flexible hours of work. Whereas policies in many countries formerly encouraged early retirement, for instance to improve employment prospects for younger workers, the OECD (2000: 13) reported that: "the unambiguous thrust of policy is now to encourage later and, in some cases, more gradual transitions from work to retirement." Higher age pensions or a one-off bonus payment are types of incentives for people to delay retirement. Increasingly there will be a need for choices, not just between work and leisure, but between different combinations of the two, providing varied ways of maintaining productivity for longer and achieving phased retirement. Sometimes taxation levels and means testing of pensions discourage older people's engagement in part-time work because this can lead to withdrawal of part or all of the age pension.

Even in Australia, a relatively 'low risk, high resilience' country (see Chap. 17), population aging is recognized as presenting significant challenges for management of the economy. In 2010 more than a quarter of Australian government spending was directed to health, age pensions and aged care. Unless spending growth is curtailed, by 2050 the share could expand to almost half. The main area of growth is spending on health, which as a proportion of GDP is projected to rise from 4.0%, near the OECD average, to 7.1% – because of aging, growing demand for health services and new technologies. The corresponding projected increase for age-related pensions is from 2.7% to 3.9% of GDP, while for aged care it is from 0.8 to 1.8 (Swan 2010: 47 and 49). Nonetheless age pensions and aged care are significant factors in higher government expenditure.

Australian strategies to improve fiscal sustainability are concerned with containing expenditure and boosting economic growth. The approaches to containing expenditure include greater targeting of pension payments through changes to income testing, raising the pensionable age, encouraging private saving for retirement income and restricting the list of free pharmaceuticals. Starting in 2017, the Age Pension age for men and women in Australia will increase from 65 to 67 years, at a rate of 6 months every 2 years. This means that persons born in 1957 and later years will not be eligible for the pension until they turn 67. Whereas many other OECD countries pay age pensions based on pre-retirement individual earnings, Australia's age pension is means-tested to target poverty alleviation. In conjunction with a higher pensionable age, this means that although Australia's projected increase in expenditure on the age pension is considerable, it is low compared with that of most other OECD countries (Swan 2010: 61). Governments have also sought to promote healthy aging and early identification and treatment of illnesses, although the fiscal benefits are uncertain. The scope and complexity of the economic issues that population aging is creating in Australia emphasizes the potential severity of its economic impact in countries that have higher levels of aging and fewer resources.

Past experience of stringency in the funding of pensions, health services and aged care, even in wealthy countries in prosperous times, inspires little confidence that living standards of old and young will be improved or maintained without major reforms. These may extend even to whole economies: "Fixing the expenditure consequences of population aging may be as much about fixing economic growth and labour markets as it is about pension reform and expenditure cutbacks" (Castles 2004: 120). Demographic pressures for spending on aged care will grow considerably in coming decades. For example, the number of persons aged 85 and over is expected to more than double in the United Kingdom between 2010 and 2050, while trebling in the United States and quadrupling in Australia. The higher figures reflect the impact of prolonged baby booms after the Second World War and high levels of immigration. As discussed in earlier chapters, aging in place and care at home are becoming dominant themes in current approaches to aged care but many differences persist. Trends towards lower rates of institutionalization have meant that nursing home accommodation is increasingly reserved for people needing 24 hour care (United Nations 2005: 55). Figures of 4–6% in long-term institutional care are most common – including in countries which also provide high levels of aged care at home through domiciliary services.

Among the different care regimes in industrial countries, Mediterranean countries form a distinctive cluster "where management of care is overwhelmingly entrusted to the family" (Bettio et al. 2006: 271). Italy has the lowest percentage of the aged in long-term care – 1.5%. Family care for dependent persons with severe disabilities in Italy has been supplemented by cash benefits that have enabled even relatively low income families to employ female migrant workers. In 2003, the benefit amounted to €436 per month and was granted to 5.8% of persons over 65. The origins of the migrant workers included the Philippines, Eastern Europe, South America and North Africa. In Rome, Naples, Milan and some other parts of Italy, the traditional family model of care is becoming a "migrant in the family" model of care (ibid.: 272). However, Italy's "do-it-yourself care by families and the private market" system lacks public regulation (Pavolini and Ranci 2008: 254). In a changing and diverse age care regime the potential for unmet need and abuse of clients increases as does mistreatment and exploitation of employees, for instance when they are untrained or on short-term contracts. Regulation of age care and service delivery should be a concomitant of all developments. Finally, a study of changes in the prevalence of severe disabilities in 12 OECD countries concluded that, despite some declines, population aging and greater longevity were leading to increasing numbers of older people with severe disabilities and in need of long-term care (Lafortune et al. 2007: 4).

16.6 Third Sector Development

Growing expenditure on pensions and other support for the aged has been hailed as "a victory for the family" in Western societies because it has shifted financial responsibility from the family to the state. Yet the victory is incomplete because health systems are under strain, some pension systems provide no more than a subsistence income, and saving for retirement is again becoming an individual and family responsibility. Moreover, future cohorts of the aged in developed countries will include many who are never married, childless or divorced – thereby weakening the influence of intergenerational support and family-based altruism. Given the varied changes that lie ahead, a priority will be to avoid unnecessary public expenditure through targeting outlays where the need is greatest. Without a strong economy, the material expectations of developed countries cannot be met, nor can provision for the vulnerable. In developing countries there is little prospect that, in coming decades, economic growth will be sufficient to enable adequate support for older people and their families. Even together, families, communities and governments cannot now provide all necessary means to adapt to the challenges of population aging. Support for families and older people is failing in developing countries generally, as well as in parts of Europe (see Chap. 17).

In these circumstances a potentially major contributor to the welfare of families and the aged is the Third Sector (Box 16.1). Expanded Third Sector involvement, implemented with international assistance in low income countries, could contribute much towards supporting community effort and securing individual and family welfare in situations of poverty and hardship. The Third Sector is part of the middle or 'third way' between sole reliance on either the market or the state to address public problems (Giddens cited by Anheier 2009: 1084). In developing countries Third Sector organizations or NGOs (non-governmental organizations) also enable aid agencies to avoid frustrations associated with the top-down policies of corrupt or inefficient governments. Moreover, the Third Sector's civil society functions – fostering participation, social integration, social capital formation and community building – are pillars of social and economic development (Anheier 2009: 1084–1085). Key strengths of the Third Sector are its willingness and ability to address neglected social concerns and obtain funding through cooperation with governments and international bodies, or through its own fund-raising expertise.

Box 16.1 Characteristics of the Third Sector

The concept of the 'Third Sector' originated from the United Nation's System of National Accounts, first devised in 1952. This identified four sectors of society: the government sector, the for-profit or business sector, the non-profit or Third Sector, and the household sector. The Third Sector is the primary partner with government in provision of community services, promotion of culture and the arts, and advocacy of civil society (Kramer 2000: 1). The Third Sector includes a diverse array of clubs and associations – e.g. concerned with sports, hobbies, and the arts – together with emergency services such as volunteer fire fighting, advocacy groups and religious, charitable and humanitarian organizations providing social services. As the number and types of Third Sector organizations has grown rapidly since the 1970s so too competition for funding has increased, leading some to undertake commercial, for profit activities to support their goals. This has caused a blurring of the boundaries with other sectors and the emergence of a "a new mixed social economy" (ibid.: 3). Aged Concern in the UK, the National Council on Aging (NCOA) in the United States and the Council on the Ageing in Australia are examples of Third Sector organizations concerned specifically with the needs of older people, although many other advocacy groups and charities also serve their interests.

In OECD countries Third Sector organizations have accounted for around 6% of total employment or nearly 10% if volunteer work is included. The sector's main sources of income vary between countries, but overall 49% is public funding, 42% derives from fees and other commercial income and 9% from the private philanthropy of foundations, corporations and individuals. Rapid growth of the sector in developed market economies is linked (i) with government social welfare spending and higher demands for services, (ii) with growing doubts about the capacity of the state to deal with a range of problems, and (iii) with the recognition that Third Sector organizations are effective providers of services. In essence, non-profits add to "the problem-solving capacity of modern societies" (Anheier 2004: 111–114 and 131). Governments retain substantial discretion over Third Sector spending through determining the direct allocation of public funds and the tax deductibility of individual and corporate donations (Wagner 1990: 304). In the United States, Germany, France, the United Kingdom and Australia a prominent feature of welfare provision is the collaboration between governments and the Third Sector. Where no such partnership exists, as in Japan and Italy, the non-profit sector is smaller in scale (Anheier 2004: 114–115). Outsourcing of social care services in the UK has expanded since 1979, such that by 2005, more than half of the social care hours in the United Kingdom were provided through Third Sector organizations (Cunningham and James 2009: 363–4). Volunteering in the UK has been worth about £38 billion per year and registered charities numbered more than 164,000 in 2005 (Shahani 2009: 3).

Third Sector development is consistent with ideas about the aged as a resource for society and social and civic engagement as personally beneficial. Although the aged are major recipients of Third Sector services and support, they also figure prominently as volunteers and donors. In the future, substantial involvement of Third Age people in Third Sector organizations could contribute much towards moderating concerns about meeting needs for service provision, as well as facilitating social capital building through linking older people with others in their own neighbourhood or community. Social welfare and health-related charities (e.g. the Salvation Army, the Red Cross, the Heart Foundation) are among the best known Third Sector organizations. They and many others have a role to play not only in addressing social concerns, but also in facilitating social network formation for older people and enhancing their well-being through friendships and many forms of social participation. Some major types of organizations, however, have declining membership, including churches and service clubs. At the same time, the nature of volunteering is changing as market forces impel non-profit organizations to become more like the for-profit sector, giving greater attention to competitiveness, efficiency and regulatory requirements including risk management and volunteer and client protection. This can break down distinctions between paid staff and volunteers as skills and qualifications take precedence, discouraging volunteers who want less formal and less demanding roles (Warburton and McDonald 2009).

Thus, although the Third Sector is gaining increasing prominence there are both positive and negative changes affecting older people's engagement with it. Other obstacles to Third Sector development include lack of security and stability of financial grants to non-profit organizations, inadequate funds to pay staff (Cunningham and James 2009) and problems of volunteer recruitment and turnover. More fundamental are differences concerning: (i) whether the government should be the main provider of social welfare, as in Sweden; (ii) whether the Third Sector should confine itself mainly to the business of service delivery rather than to associated civil society functions, as in France and Germany; or (iii) whether large scale provision of social welfare services has any priority at all if neither the state nor the non-profit sector have major roles, as in Japan (Anheier 2004: 124–7).

Existing differences between countries in their approaches to social welfare provision are likely to long maintain contrasts in the development of the Third Sector as a service provider. Some high income countries such as the United States, Canada, the UK and Australia, are already ahead here with high levels of participation in non-profit organizations and substantial contributions to the economy ensuing. At the international level there were over 47,000 NGOs in 2001, many of which had formal organizational links with the United Nations Development Program (UNDP), the World Health Organization and the World Bank. Some examples are Amnesty International with more than one million members and affiliates, the Friends of the Earth with around 5,000 local groups and one million members, and Care International which has over 10,000 professional staff (Anheier 2004: 119). International NGOs are primarily Western or of Western origin (Miller 2007: 353). Many NGOs participated in the Second World Assembly on Ageing. Their prominence in the deliberations is a sign of the expansion of the role of non-profit organizations in supporting international, national and local welfare initiatives for the aged.

16.7 Conclusion

Although responses to population aging need to accord with national circumstances, population aging is a global phenomenon about which more can be learnt through collaboration between countries than through considering individual countries in isolation. International cooperation has a major role in conducting policy-related research and in supporting welfare-related programs in low income countries. Increasingly too the international setting is influencing quality of life in aging societies. Globalization directly affects individual and population aging through its impact on international economic linkages, the diffusion of Western values in contemporary life styles, and the spread of medical knowledge of life-preserving strategies. At the same time, global environmental concerns are becoming an increasingly significant aspect of the international setting and are likely to compete with aging for attention and funding. Population aging is occurring in a period of critical decisions concerning climate change and conservation of resources and the environment, all of which require timely action because the trends are cumulative and far-reaching.

Research on aging, as well as research in the social sciences generally, is placing greater emphasis on longitudinal studies because of their potential to yield new insights and better explanations. Investigations of health in later life, for example, are using this approach to provide data on risk factors and other influences on health from early years (Buys and Anstey 2008). Butz and Torrey (2006) referred to longitudinal studies as one of the means of meeting "the fundamental challenge in the social sciences" of moving from complicated correlations to useful prediction. Other strategies they identified include improved statistical methods, geographic information systems, international replication, cross-disciplinary inquiry, and microsimulation modelling. Also needed are in-depth qualitative research and a greater comparative understanding of policy and practice (Harper 2006: 118). All of these reflect the extensive nature of the agenda for the study of aging and the aged. The scope for innovation in research, however, is not a reason for delaying action on population aging as broad trends and their consequences are already well recognized. Readiness to plan for the future of population aging will itself be a stimulus for research and advances in knowledge.

References

Anheier, H. (2004). Third sector-third way: Comparative perspectives and policy perspectives. In J. Lewis & R. Surender (Eds.), *Welfare state change: Towards a third way?* (pp. 111–134). Oxford: Oxford University Press.

Anheier, H. (2009). What kind of nonprofit sector, what kind of society?: Comparative policy reflections. *American Behavioral Scientist, 52*(7), 1082–1094.

Bettio, F., Simonazzi, A., & Villa, P. (2006). Change in care regimes and female migration: The "care drain" in the Mediterranean. *Journal of European Social Policy, 16*(3), 271–285.

Butz, W. P., & Torrey, B. B. (2006). Perspective: Some frontiers in social science. *Science, 312*(5782), 1898–1900.

Buys, L., & Anstey, K. (2008). Theoretical, empirical and policy inputs to modelling healthy ageing. *Dialogue (Academy of the Social Sciences in Australia), 27*(1), 70–71.

Castles, F. G. (2004). *The future of the welfare state: Crisis myths and crisis realities*. Oxford: Oxford University Press.

Coleman, D. A. (1998). *Reproduction and survival in an unknown world* (NIDI Hofstee lecture series, 15). The Hague: Netherlands Interdisciplinary Demographic Institute.

Coleman, L. (2008). Family policy: Framework and challenges. In F. Coulmas, H. Conrad, A. Schad-Seifert, & G. Vogt (Eds.), *The demographic challenge: A handbook about Japan* (pp. 749–763). Leiden/Boston: Brill.

Coleman, D., & Rowthorn, R. (2011). Who's afraid of population decline? A critical examination of its consequences. *Population and Development Review, 37*(Supplement), 217–248.

Cortazzi, H. (2002, April). Greying nation. *The World Today, 58*(4), 11–12.

Costello, P. (2002, August 07). *The paths to increasing Australian prosperity, address to Australian Financial Review leaders' luncheon*, Sydney. http://www.treasurer.gov.au/DisplayDocs.aspx?pageID=&doc=speeches/2002/004.htm&min=phc. Accessed Mar 2009.

Coulmas, F. (2008). *Population decline and ageing in Japan: The social consequences*. London/New York: Routledge.

Cunningham, I., & James, P. (2009). The outsourcing of social care in Britain: What does it mean for voluntary sector workers? *Work, Employment and Society, 23*(2), 363–375.

Dorius, S. F. (2008). Global demographic convergence? A reconsideration of changing intercountry inequality in fertility. *Population and Development Review, 34*(3), 519–537.

Harper, S. (2006). *Ageing societies: Myths, challenges and opportunities*. London: Arnold.

Heigl, A. (2004). In Review of J. B. Kune (Ed.), *On global ageing: Old-age income systems in the EU and other major parts of the world*. Heidelberg: Physica. *European Journal of Population 20*, 287–292.

Jones, G. W. (1993). Is demographic uniformity inevitable? *Journal of the Australian Population Association, 10*(1), 1–16.

Kaneko, R. (2008). Statistical foundations of population projections. In F. Coulmas, H. Conrad, A. Schad-Seifert, & G. Vogt (Eds.), *The demographic challenge: A handbook about Japan* (pp. 41–59). Leiden/Boston: Brill.

Kramer, R. M. (2000). A third sector in the third millennium? *Voluntas: International Journal of Voluntary and Nonprofit Organizations, 11*(1), 1–23.

Lafortune, G., Balestat, G., & The Disability Study Expert Group Members. (2007). *Trends in severe disability among elderly people: Assessing the evidence in 12 OECD countries and future implications* (OECD Health Working Papers, No. 26). Paris: OECD.

Lindh, T. (2005). Productivity is a system property and need not decrease with the age of the workforce. *Vienna Yearbook of Population Research 2005*, 7–9.

MacKellar, F. L. (2003). The predicament of population ageing: A review essay. *Vienna Yearbook of Population Research 2003*, 73–99.

McDonald, P., & Kippen, R. (2001). Labor supply prospects in 16 developed countries, 2000–2050. *Population and Development Review, 27*, 1–32.

Miller, J. (2007). Neither state nor market: NGOs and the international third sector. *Global Media and Communication, 3*, 352–355.

OECD. (1998, June). *Maintaining prosperity in an ageing society: Summary*. http://www.oecd.org/dataoecd/21/10/2430300.pdf. Accessed Sept 2010.

OECD. (2000). *Reforms for an ageing society*. Paris: Organisation for Economic Co-operation and Development.

OECD. (2010). *Restoring fiscal sustainability: Lessons for the public sector*. Paris: Organisation for Economic Co-operation and Development. http://www.oecd.org/dataoecd/1/60/44473800.pdf. Accessed June 2011.

Ogawa, N. (2008). Population ageing and economic growth: The role of the two demographic dividends in Japan. In F. Coulmas, H. Conrad, A. Schad-Seifert, & G. Vogt (Eds.), *The demographic challenge: A handbook about Japan* (pp. 821–839). Leiden/Boston: Brill.

Olschleger, H. D. (2008). Fertility and mortality. In F. Coulmas, H. Conrad, A. Schad-Seifert, & G. Vogt (Eds.), *The demographic challenge: A handbook about Japan* (pp. 25–40). Leiden/Boston: Brill.

Pavolini, E., & Ranci, C. (2008). Restructuring the welfare state: Reforms in long-term care in Western European countries. *Journal of European Social Policy, 18*(3), 246–259.

Pozzebon, S. (2004). *The future of pensions in Canada*. Pension Research Council Working Paper 2004–15. Philadelphia: The Wharton School, University of Pennsylvania. http://prc.wharton.upenn.edu/prc/PRC/WP/WP2004-15.pdf. Accessed Jan 2010.

Prskawetz, A. (2005). Will population ageing decrease productivity? *Vienna Yearbook of Population Research 2005*, 1–3.

Roberts, G. S. (2008). Immigration policy: Framework and challenges. In F. Coulmas, H. Conrad, A. Schad-Seifert, & G. Vogt (Eds.), *The demographic challenge: A handbook about Japan* (pp. 765–779). Leiden/Boston: Brill.

Shahani, V. (2009). *The contribution of volunteers to community cohesion in Sydney*. http://www.wcmt.org.uk/reports/601_1.pdf. Accessed May 2011.

Skirbekk, V. (2005). Population ageing negatively affects productivity. *Vienna Yearbook of Population Research 2005*, 5–6.

Swan, W. (2010). *Australia to 2050: Future challenges* (The 2010 intergenerational report). Canberra: Commonwealth of Australia.

Tobback, B. (2005, October 17–18). Chair's conclusions. *High-level forum on ageing and employment policies*, Brussels. www.oecd.org/olderworkersforum. Accessed Apr 2010.

United Nations. (2004). *Policy responses to population decline and ageing* (Population bulletin of the United Nations, special issue nos. 44/45). New York: Department of Economic and Social Affairs, Population Division.

United Nations. (2005). *Living arrangements of older persons around the world*. United Nations Department of Economic and Social Affairs, Population Division. New York: United Nations, ST/ESA/SER.A/240.

United Nations. (2009a). *World population prospects: The 2008 revision population database*. New York: Population Division of the Department of Economic and Social Affairs. http://esa.un.org/UNPP/index.asp?panel=2. Accessed Apr 2010.

United Nations. (2009b, December). *World population ageing 2009*. New York: United Nations Department of Economic and Social Affairs, Population Division, ESA/P/WP/212. http://www.un.org/esa/population/publications/WPA2009/WPA2009_WorkingPaper.pdf. Accessed Oct 2010.

United Nations. (2010). *World population policies 2009*. New York: Department of Economic and Social Affairs, Population Division.

Wagner, A. (1990). The nonprofit sector in Switzerland: Taxonomy and dimensions. In H. K. Anheier & W. Seibel (Eds.), *The third sector: Comparative studies of nonprofit organizations* (pp. 303–312). Berlin/New York: Walter de Gruyter.

Warburton, J., & McDonald, C. (2009). The challenges of the new institutional environment: An Australian case study of older volunteers in the contemporary non-profit sector. *Ageing and Society, 29*, 823–840.

Wilson, C. (2001). On the scale of global demographic convergence 1950–2000. *Population and Development Review, 27*, 155–171.

Wilson, C. (2003). Homeostasis. In P. Demeny & G. McNicoll (Eds.), *Encyclopedia of population* (pp. 493–496). New York: Macmillan Reference.

Wilson, C., & Airey, P. (1999). How can a homeostatic perspective enhance demographic transition theory. *Population Studies, 53*, 117–128.

Chapter 17
Risks and Resilience

The trend towards population ageing is bringing about profound changes for all generations and most areas of economic and social activity. As Europe enters the twenty-first century, the demographic trend towards an ageing society is becoming a highly important issue for social protection, the labour market, politics, design and technology, education and culture, in short for the whole economy and all of society.

(Walker 1999)

17.1 Crisis or Business as Usual?

Like the 'inconvenient truth' of global warming, national and global population aging raise long-term issues for the twenty-first century. Demographic transition theory formerly portrayed population aging as a fairly uniform, gradual and benign process. This supports a 'business as usual' perspective on the prospects for aging societies. Continuing changes in societies, however, have shown population aging to be more diverse, rapid and problematic, sometimes provoking fears of impending crises. Earlier chapters have examined these differences in relation to a range of subject areas. This final chapter compares selected countries in terms of indicators of their exposure to 'risks' from population aging and their resilience to negative consequences. In doing so, the chapter summarizes the outlook for population aging in countries with the oldest or largest aged populations and evaluates the business as usual and crisis perspectives.

17.2 Comparing Risks

The analysis of risks focuses on the 38 countries used for comparisons in earlier chapters. These are 34 countries, with populations of one million or more, that had the oldest or largest aged populations, together with four others of particular interest

D.T. Rowland, *Population Aging: The Transformation of Societies*,
International Perspectives on Aging 3, DOI 10.1007/978-94-007-4050-1_17,

(see Preface). No single measure of aging represents all aspects of the process, hence the use of a set of indicators relevant to a range of different countries. These include high positive or negative momentum, low fertility, rapid growth in the older ages, labour force decline, high percentages in older ages and high aged dependency ratios. Most of the indicators were calculated from estimates and medium variant projections in the 2008 Revision of the U.N.'s *Population Trends and Prospects*. Future prospects, in terms of what seemed plausible in 2008, are important because the major changes in aging populations lie ahead, and because projections can serve as forewarnings of adverse developments. The long-range implications of current changes can inform policy decisions because it may take decades for some initiatives to become effective. For example, cohort self-sufficiency – in relation to retirement income and funding of health services and long term care – needs to be built over the whole of a cohort's working life. Thus some of the indicators of aging and risk are based on projections to 2050, not because the figures are taken as predictions, but because they illustrate long run consequences of current trends.

As in the classification of age structures in Chap. 3, cluster analysis provided a suitable means of grouping countries – this time in terms of their indicators of risk. The cluster analysis identified groups of countries with similar characteristics. There were 2–8 countries in each group. Most consisted of contiguous countries – indicating regional similarities. Indonesia and Norway were exceptions in being grouped with distant places. Japan, China and Pakistan were ungrouped because they had distinctive characteristics. They would have remained separate even after combining some of the existing groups. The cluster analysis revealed the main patterns of variation in the mix of statistical indicators in different countries, highlighting differences in national experiences of population aging. In the analysis the model population serves as a benchmark for comparisons (see Fig. 16.1). A demographically favourable long term outcome from population aging would be an age structure similar to that of the model population, which has a relatively low percentage in the older ages (22.5%) and a relatively high percentage in the working ages (59.3%). Any long-lived society moving towards population stabilization would develop an age structure similar to that of the model. A number of developed countries are favourably positioned to achieve an approximation to it by mid-century, after population aging has run its course.

There were three main patterns in the risk factors:

- A developed country pattern associated with a trend towards high proportions in the older ages, due to low fertility. Among the countries with this pattern, risks from aging ranged from relatively low (groups 1 and 2) to relatively high (groups 3–5 and Japan).
- A developing country pattern characterized by very high growth in the numbers of the aged, due to relatively high birth rates. Numerical growth represented a substantial risk for countries conforming to this broad pattern (groups 7–8 and Pakistan).
- An intermediate pattern in the three high growth Asian economies – Hong Kong and Singapore (group 6) together with China. These countries had a combination of the characteristics of the developed and developing country patterns, namely low birth rates and high growth in their numbers and percentages aged 65 and over.

Within these three broad patterns there were important differences, as shown in Table 17.1. The table presents the average characteristics of each group and its constituent countries, together with the characteristics of the ungrouped cases. It also includes the characteristics of the model population. Figure 17.1, which depicts the group membership of countries in Europe in terms of the risk analysis, reveals quite marked regional patterns. The regional names correspond broadly with the United Nations' regions of Europe. Exceptions are Germany and Austria: while the UN places them in Western Europe, they are referred to here as parts of Central Europe, as they are in some other sources. Similarly Poland and the Czech Republic, sometimes described as parts of Central Europe, are in a group consisting of countries in the western part of Eastern Europe (group 4, Eastern Europe I).

17.3 Developed Countries with Relatively Low Risks

The nearer countries converge to the characteristics of the model population, the less the demographic challenges in adapting to population aging. In a number of ways group 1 – North America and Australasia – was closest to the model population. Its fertility (2000–2005) was nearest to the model's replacement fertility level, and its dependency ratios and percentage aged 65 and over in 2050 were only slightly above the model's figures. Where the group departed from the model was in the relatively high projected growth of its labour force and aged population, associated with the aging of large baby boom generations and the effects of net immigration, which boosts numbers in the younger working ages. The age structures for the United States in 2010 and 2050 illustrate these developments (Fig. 17.2). In 2010 the group's momentum was only 5%, positioning it well in relation to achieving population stabilization. Norway was placed in group 1 mainly because, like the United States and Australia, it had higher projected growth in the labour force and older ages than did other Nordic countries.

The other group with favourable characteristics overall consisted of the eight countries from Western and Northern Europe (group 2). With an average TFR of 1.7 and population momentum of −5% in 2010, the data are indicative of populations with a prospect of avoiding hyper-aging and stabilizing population numbers and age structures. As in the first group, population aging in Western and Northern Europe is reflected in rising dependency ratios and older people eventually far outnumbering children, although the changes are generally more moderate than in other European countries. Projections show labour force numbers remaining reasonably stable in the long term and the proportion in the older ages reaching little more than 25%. In this group the projected changes associated with population aging are less disruptive than elsewhere in Europe. This is probably a legacy of a protracted pace of population change through time, without extremes of growth or decline. The age pyramids for France illustrate a concentration of growth in the older ages, rectangularization of the age structure and convergence towards the model population (Fig. 17.2). Indices of dissimilarity show that all of the populations in the first two groups had age structures that differed from the model age structure by between 6% and 10% in 2010 and only 3–5% in 2050.

Table 17.1 Grouping of countries in terms of indicators of population aging, 2000–2050

Groups and ungrouped countries	Female population momentum 2010	Average total fertility rate 2000–2005	Life expectancy at birth, both sexes 2006	% aged 65 and over 2025	% aged 65 and over 2050	% change in the total population 2010–2050	% change in the labour force ages (15–64) 2010–2050	% change in ages 65 and over 2010–2050	Total dependency ratios 2050	Aged dependency ratios 2050	Aging index 2050 (aged per 100 children)
Developed Country Pattern											
(a) Relatively Low Risk Countries											
North America & Australasia (Group 1)											
United States	6.7	2.04	78	18.1	21.6	27.2	16.8	111.7	62.9	35.1	126.5
Canada	–1.8	1.52	81	20.5	25.5	31.0	10.9	136.9	69.9	43.4	163.7
Australia	8.9	1.75	82	19.1	23.8	33.5	18.3	128.5	67.9	39.9	142.8
New Zealand	10.8	1.95	80	18.1	23.2	24.3	12.1	122.0	66.0	38.5	140.2
Norway	2.3	1.80	80	19.4	23.8	22.5	10.8	94.1	66.9	39.8	146.4
Group averages	5.4	1.81	80	19.0	23.6	27.7	13.8	118.6	66.7	39.3	143.9
Western and Northern Europe (Group 2)											
United Kingdom	–4.4	1.70	79	19.4	22.9	16.9	7.5	61.2	64.7	37.7	139.3
France	0.4	1.88	81	22.6	26.9	8.0	–5.0	71.4	75.9	47.3	165.7
Switzerland	–8.4	1.42	82	21.9	26.0	12.0	–3.3	69.0	71.5	44.7	166.1
Belgium	–6.5	1.64	79	22.2	26.6	7.4	–6.4	64.1	74.3	46.4	165.9
Netherlands	–3.9	1.73	80	21.7	25.6	4.5	–8.7	74.2	70.7	43.7	161.9
Denmark	–7.9	1.76	79	21.3	23.8	1.3	–6.8	44.5	66.5	39.6	147.8
Sweden	–3.7	1.67	81	21.7	24.1	13.7	3.6	49.5	68.3	40.5	146.0
Finland	–6.2	1.75	79	23.9	25.9	1.8	–10.4	53.3	71.6	44.4	163.8
Group averages	–5.1	1.69	80	21.8	25.2	8.2	–3.7	60.9	70.4	43.0	157.1
(b) Relatively High Risk Countries											
Central and Southern Europe (Group 3)											
Germany	–18.7	1.35	80	25.1	32.5	–14.1	–28.7	36.3	82.0	59.1	258.4
Austria	–10.9	1.39	80	22.0	29.4	1.5	–15.2	70.1	76.7	52.0	209.9

Italy	−19.9	1.26	81	24.4	33.3	−5.0	−22.6	54.5	87.7	62.4	246.8
Greece	−18.5	1.28	80	22.4	31.3	−2.2	−20.1	67.2	81.6	56.8	229.9
Slovenia	−17.1	1.23	78	22.4	30.2	−3.4	−23.1	77.5	80.1	54.4	211.8
Spain	−13.7	1.29	81	20.4	31.8	13.1	−11.0	109.5	87.1	59.5	215.4
Portugal	−13.5	1.44	79	22.4	32.1	−6.7	−23.9	67.8	83.0	58.8	242.4
Group averages	−16.0	1.32	80	22.7	31.5	−2.4	−20.6	69.0	82.6	57.6	230.6
Eastern Europe I (Group 4)											
Poland	−5.3	1.25	75	21.0	29.9	−15.8	−32.7	86.5	74.4	52.2	234.9
Czech Republic	−13.8	1.19	77	20.5	27.6	−1.1	−20.2	78.8	75.4	48.4	178.9
Croatia	−13.8	1.36	76	22.2	28.2	−13.2	−26.6	41.4	74.5	49.2	195.3
Bulgaria	−21.6	1.25	73	21.9	30.3	−28.1	−42.3	23.9	80.8	54.8	211.6
Group averages	−13.6	1.26	75	21.4	29.0	−14.6	−30.5	57.7	76.3	51.2	205.2
Eastern Europe II (Group 5)											
Russia	−13.2	1.30	66	17.7	23.4	−17.3	−30.8	50.2	65.7	38.8	144.7
Ukraine	−16.3	1.15	67	18.6	24.7	−22.9	−35.2	22.2	68.6	41.6	154.0
Hungary	−17.1	1.30	73	20.3	26.2	−10.4	−23.2	43.2	69.3	44.3	177.2
Latvia	−15.1	1.25	71	19.5	25.9	−17.4	−29.4	22.8	70.4	44.2	168.4
Group averages	−15.4	1.25	69	19.1	25.0	−17.0	−29.6	34.6	68.5	42.2	161.1
Japan	−21.1	1.30	83	29.7	37.8	−20.0	−36.5	34.2	96.3	74.3	337.5
Intermediate Pattern											
Singapore & Hong Kong (Group 6)											
Singapore	−3.5	1.36	80	22.9	32.6	7.9	−18.2	243.6	77.8	57.9	291.3
Hong Kong S.A.R.	−16.3	0.98	82	22.1	32.6	22.0	−9.4	208.7	78.1	58.0	288.7
Group averages	−9.9	1.17	81	22.5	32.6	14.9	−13.8	226.2	78.0	58.0	290.0
China	9.2	1.77	73	13.4	23.3	4.6	−10.6	196.7	62.9	38.0	152.8

(continued)

Table 17.1 (continued)

Groups and ungrouped countries	Female population momentum 2010	Average total fertility rate 2000–2005	Life expectancy at birth, both sexes 2006	% aged 65 and over 2025	% aged 65 and over 2050	% change in the total population 2010–2050	% change in the labour force ages (15–64) 2010–2050	% change in ages 65 and over 2010–2050	Total dependency ratios 2050	Aged dependency ratios 2050	Aging index 2050 (aged per 100 children)
Developing Country Pattern											
Mexico, Brazil & Indonesia (Group 7)											
Mexico	47.0	2.40	74	10.6	22.1	16.6	9.7	293.8	62.2	35.9	136.5
Brazil	28.4	2.25	72	11.4	22.5	11.8	3.8	265.3	59.3	35.9	153.4
Indonesia	30.6	2.38	68	9.0	18.6	23.9	17.9	279.3	56.3	29.1	106.7
Group averages	35.4	2.34	71	10.3	21.1	17.4	10.4	279.5	59.3	33.6	132.2
India and Bangladesh (Group 8)											
India	36.7	3.11	63	7.3	13.7	32.9	40.7	271.4	47.0	20.2	75.5
Bangladesh	41.9	2.80	63	6.1	14.9	35.3	38.9	409.7	49.5	22.3	82.0
Group averages	39.3	2.96	63	6.7	14.3	34.1	39.8	340.6	48.3	21.3	78.7
Pakistan	55.4	4.44	63	5.1	10.0	81.4	104.5	343.9	49.6	15.0	43.2
Model Pop.	0.0	2.07	82	22.5	22.5	0.0	0.0	0.0	68.7	38.0	123.8
Data sources:	c.	a.	b.	a.	a.	a.	a.	a.	a.	a.	a.

Sources: Cluster analysis of international data from: (a) United Nations (2009); (b) WHO (2009); (c) Population Reference Bureau (2008) and West model life tables

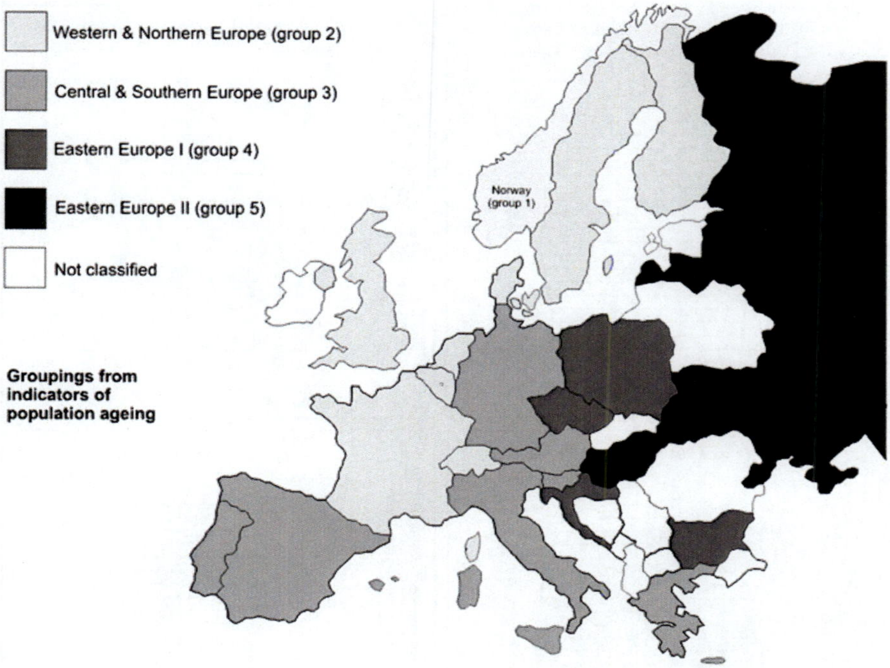

Fig. 17.1 Groups of European countries in the risk analysis (Source: Table 17.1)

17.4 Developed Countries with Higher Risks

The very low fertility rates of the countries in Central and Southern Europe (group 3) and Eastern Europe (groups 4 and 5) parallels a range of other indicators of high risk of adverse consequences from population aging. These include negative momentum, population decline, and hyper-aging. Hyper-aging is projected for mid-century in the countries of Central and Southern Europe (group 3), including Germany, Austria, Italy, Greece and Spain. Associated with this are very high aging indices – where the aged outnumber children by more than 2 to 1. In 2010, negative momentum was marked in group 3, averaging −16%. Group 3 could also experience a substantial decline in labour force numbers, although not as great as in Eastern Europe. All of these developments represent extrapolations from the situation in the first decade of the twenty-first century. Accordingly, this high risk scenario, cumulating over decades, is a forewarning rather than a prediction. The projections do not indicate any tendency towards convergence of the national age structures towards the model, as differences are relatively high in both 2010 and 2050. In these circumstances it is perhaps surprising that Austria and Italy and some other 'high risk' countries have similar population totals in 2010 and 2050. Yet only superficially is this indicative of stability because while younger age groups will decrease, older

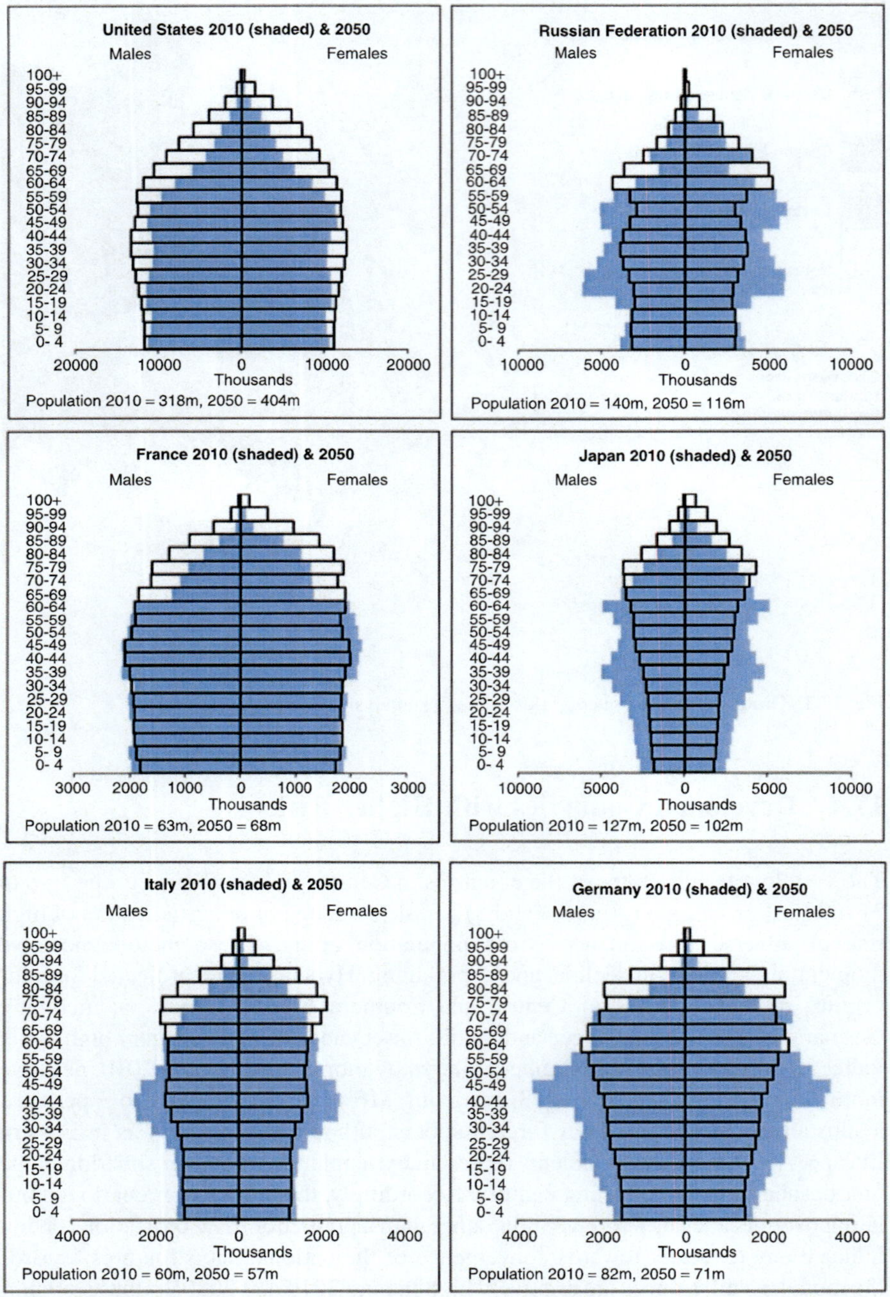

Fig. 17.2 Age structures of selected national populations, 2010 and 2050 (Source: Calculated from United Nations 2009)

age groups will increase. Unless there is higher fertility, population decline will gather pace towards mid-century as the large generations at the top of the age pyramid reach the end of their lives.

In the two Eastern European groups many factors have contributed to very low fertility including political upheaval, economic uncertainty, unemployment, declining incomes, larger gaps between rich and poor, and cutbacks in government support for families with children, such as funding for day care centres (United Nations 2004: 60). The Eastern European groups are projected to experience particularly high levels of decline in labour force numbers as well as in the population overall – only in Japan are such figures exceeded. A restraining influence on aging in parts of Eastern Europe is quite low life expectancy, especially in group 5 which includes Russia and Ukraine. In group 4, Poland and Bulgaria, which have higher life expectancies, could be approaching hyper-aging by 2050.

Among the developed countries Japan has by far the most unfavourable set of demographic indicators of risk including low fertility, early onset of hyper-aging, working age numbers declining by more than a third between 2010 and 2050 and the highest aging index in 2050 of 338 older people per 100 children. Japan also has the highest projected dependency ratios. The record level of negative momentum in Japan's age structure in 2010, namely –21%, is indicative of a steep decline in population numbers (Table 17.1). Japan's projected age structure in 2050 (Fig. 17.2) is the most dissimilar from the model age structure (16% difference), followed closely by those for Singapore and Hong Kong (15% and 14%).

17.5 High Growth Asian Economies

Singapore and Hong Kong (group 6) have a distinctive 'intermediate' pattern of aging with some of the characteristics of Japan (high negative momentum, low fertility, high life expectancy, a high aging index and hyper-aging in 2050), but with more moderate labour force decline. Singapore and Hong Kong also exhibit one the main distinguishing features of the developing country pattern, namely high projected numerical growth in the older ages 2010–2050 (Table 17.1). Similarly, China has a mix of the demographic patterns of aging in developed and developing countries. Its trend in aging particularly reflects the impact of government policies and socio-economic development on birth rates (Cai 2010). The country's birth rates underwent a remarkable decline in the 1970s, owing to the nation-wide 'Later (marriage), Fewer (children), Longer (birth intervals)' campaign. Between 1970 and 1978 the birth rate fell from 6 children per woman to 3, that is even before the launch, in 1979, of the more stringent One Child Policy (Tien et al. 1992: 10ff). China's birth rate is now similar to that of countries in Western and Northern Europe and its age structure is projected to show marked convergence to within 7% of the model profile in 2050. High growth in the older ages is counterbalancing numerical losses in the labour force ages. China's 2050 age structure also indicates a deficit of females, assuming continuing preferences for male children.

17.6 Developing Countries

The remaining countries, in groups 7 and 8, display variations on the developing country pattern, namely low percentages in older age groups now and decades hence, together with high growth in the aged population through time. Other characteristics are birth rates above replacement and high positive population momentum. Except for Pakistan, the developing countries had total fertility rates of less than 3 children per woman in 2000–2005, which would moderate their population growth were it not for the high inherent momentum in their age structures. The consequences of their past high fertility, in conjunction with longer life expectancies and fertility still above replacement, are rapid population growth overall and the prospect of great increases in the numbers at older ages. Even by mid-century some developing countries are likely to be far from completing the transition from young to old age structures.

The countries of group 7 (Brazil, Mexico and Indonesia) are moderate versions of the developing country pattern, on account of their relatively low birth rates (TFR 2.3, 2000–2005) and relatively high percentages in the older ages by 2050 (21% 65 and over). Nevertheless, like developing countries generally, they will experience major increases in the older age groups (Table 17.1). Indonesia, which had more favourable demographic characteristics than the other Asian developing countries in the data set, was grouped with Mexico and Brazil.

India and Bangladesh (Group 8) are less restrained examples of the developing country pattern with mid-century figures of around 14% aged 65 and over and the already large numbers in older ages more than three times greater than in 2010. Pakistan (ungrouped) has the most extreme form of the pattern on account of the high figures for its birth rate, momentum (55% in 2010) and labour force growth, together with the lowest percentage in the older ages at mid-century. It had the youngest of all the age structures in the data set and the highest dissimilarity from the model population in 2010 – 30%, falling to 14% in 2050 (see Fig. 14.2).

In conclusion, the indictors of risk show the extent to which aging is transforming populations. All countries face significant challenges, even those where, in a comparative setting, the changes appear more moderate. Changes in the labour force ages are particularly important because of their bearing upon economic resources. Developing countries are projected to experience overall growth in the labour force ages 2010–2050, as are North America and Australasia together with Norway, Sweden and the United Kingdom. Elsewhere, labour force decline is projected to be a major trend with the highest losses in Japan and parts of Eastern Europe. Their huge declines of 30% or more could be averted only through higher birth rates. China too is destined to experience a decline of about 11% in its labour force ages over the same period, in conjunction with almost a trebling of its aged population. Even the moderate increases in the labour force in group 1 (North America and Australasia) and the relatively small falls in group 2 (Western Europe and Northern Europe) are potential concerns given the large prospective increases in their aged populations (120% and 60% respectively 2010–2050, Table 17.1). Substantial differences between the growth rates in the older ages and the working ages are present in all countries. The challenges are greatest where there are very high rates for the former and negative rates for the latter (see Fig. 16.2). Moreover, the aging indices – the number of aged per 100 children – are a telling sign of the transformation wrought through aging, creating societies in which the aged far outnumber children. The projections anticipate that in Japan, Singapore and Hong Kong, the aged in 2050 will outnumber children by around 3 to 1, while in the Southern and Central European countries the figure could be more than 2 to 1.

17.7 Resilience

Collectively, the measures of risk summarise the extent to which aging is transforming populations. Whether countries can adapt well to population aging will depend partly on some current characteristics indicative of resilience amid prolonged demographic changes. Resilience is the ability to negotiate changes and avoid or manage negative circumstances or risks. It is a more complex subject area than demographic risks because there are many factors that can contribute to resources for addressing population aging. Also, comparable data for different countries are mostly incomplete. The extent to which countries work proactively towards preparing for future changes will influence whether risks become crises. The nature of broad policy strategies is fundamental to resilience and the quality of later life. Internationally agreed documents from the United Nations and the OECD provide guidelines for national policies. The range of recommended policy initiatives is very large and there is no single agenda appropriate for all countries. This section focuses on information about economic resources, human capital resources, and policy initiatives.

17.7.1 Economic Resources

Population aging has major implications for national economies because it leads to greater demands on government expenditure and, in many cases, declining revenue from taxation. Although economic factors provide only partial coverage of the prerequisites for successfully negotiating future developments, they are among the most important. Unfavourable demographic conditions are more likely to become a predicament for low income countries than for more prosperous ones. GDP per capita – based on the total market value of goods and services produced within a country – is a measure of material living standards and, hence, of resilience. High GDP generally denotes national prosperity and the presence of resources for social expenditure. It does not take account of the benefits to the economy of volunteering and unpaid work, both of which could become more prevalent as populations age. The data here refer to GDP per capita expressed in American dollars and purchasing

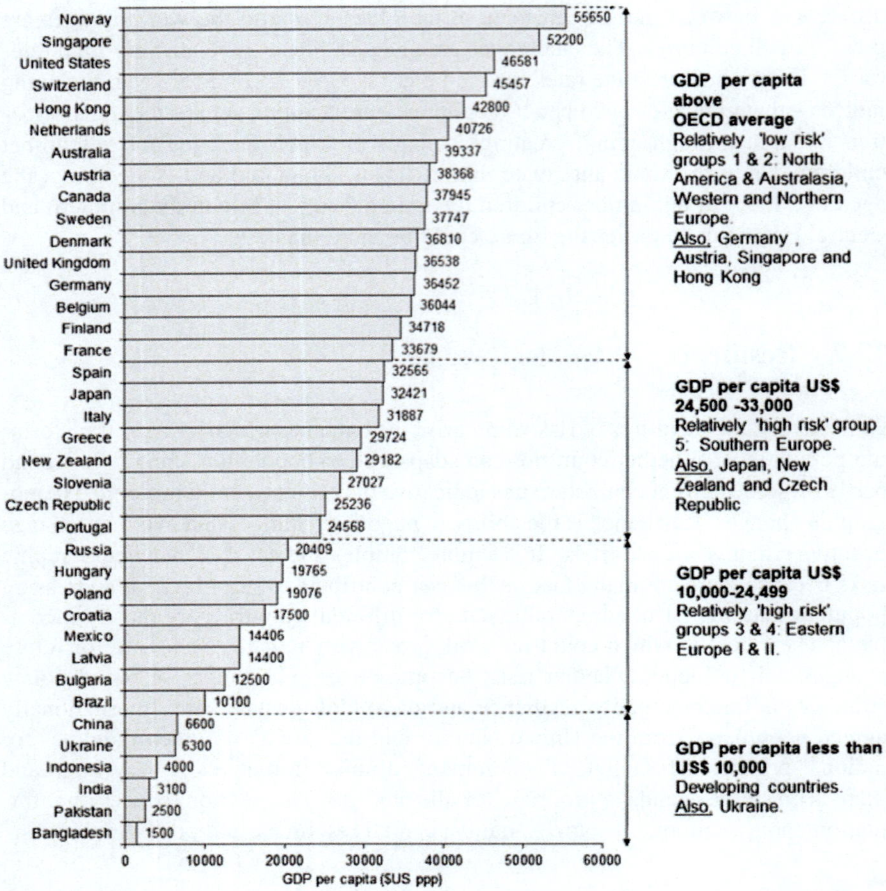

Fig. 17.3 GDP per capita ($US ppp) 2009, selected countries (Sources: OECD 2009; CIA 2010)

power parities (ppp), which equalize the purchasing power of different currencies (OECD 2010: 32).

Figure 17.3 depicts GDP per capita for the 38 countries in the risk analysis. The main feature is that the grouping of countries in terms of their demographic risks aligns broadly with their GDPs. Thus the developed countries with relatively low risks (groups 1 and 2, Table 17.1) are in the top section of the chart where GDP per capita equals or exceeds the OECD average of US$33,697. Also located there are Germany and Austria (group 3) together with the high growth Asian economies of Singapore and Hong Kong (group 6). Next, with incomes between US$25,000 and 33,000, are the Southern European countries of Group 3, described earlier as having relatively high risk. Japan and New Zealand are also in this lower income range. Further down the chart, with GDP per capita of US$10,000–24,999, are countries in the two high risk Eastern European groups (groups 4 and 5). The developing countries of Mexico and Brazil are in the same income bracket as most of the Eastern European countries. The other developing countries, including China, have the lowest incomes. China's GDP per capita was only 20% (US$6,600) of the OECD average in 2009. China, has a high total GDP, but on a per capita basis it is low. The overall pattern suggests that relatively low demographic risk and relatively high economic resilience go together. Contradicting this pattern are Austria, Germany, Singapore and Hong Kong – with high risk and high resilience – and New Zealand with low risk and low resilience, at least on this single indicator. The global financial and economic crisis, however, weakened the resilience even of high income countries, as unemployment and public and private debt grew (see World Bank 2011).

Other economic indicators of resilience are the two components of GDP – labour productivity (GDP per hour worked) and labour utilization (average hours worked per person in the whole population). GDP per capita is the product of these two components; labour utilization combines the effects of changes in population numbers and labour force participation rates (Swan 2010: 16–17). Labour utilization and labour productivity are key aspects of economic resilience and attention to both is necessary in addressing economic concerns of aging populations. Advantages in one component of GDP can offset disadvantages in the other. For example, labour utilization in France was 20% below the OECD average but the country's labour force productivity enabled it to achieve a GDP close to the OECD average. Levels of labour productivity in 2009 appear to have varied in a similar pattern to GDP per capita. Countries in the relatively low risk groups (1 and 2) mostly had the higher levels of labour productivity above the OECD average of US$43 per hour worked. Norway, with its earnings from oil and natural gas had by far the highest level of labour productivity (US$73 per hour worked) followed by the United States (US$57). Germany, Austria and Spain also had above average productivity. Most Southern European countries, together with Japan and New Zealand, generally had productivity below the OECD average. New Zealand's low labour productivity is probably due partly to immigrant labour force growth in low-skilled occupations, lack of economies of scale, and distance from international markets. The few Eastern European countries for which data were available had the lowest productivity. Thus the available productivity figures broadly parallel the pattern in the risk data, but with a number of exceptions including Germany, Austria and New Zealand. Data on labour productivity were unavailable for 11 of the 38 countries, including most of the developing countries and countries in Eastern Europe, together with Hong Kong and Singapore.

Greater labour utilization can potentially offset lower levels of labour productivity if more hours are worked per head of population. Participation depends on the proportion of people of working age who want paid employment, the average hours worked and the unemployment rate. Low levels of utilization can occur if there is a high proportion of children in the population, or if many of working age are unemployed or not in the labour force. While some countries, such as Switzerland achieved a high GDP through high labour utilization, others, such as the Russian Federation and the Czech Republic had low GDPs despite labour utilization well above the OECD average. Overall, the available data on labour utilization do not indicate consistent regional differences. More than half the countries were within 10% of the OECD average. Also, the 2009 figures are not representative of long-run developments because of the impact of the Global Financial Crisis (GFC).

In some countries, labour utilization formerly benefitted from the maturing of baby boom generations and rising labour force participation of women, but the effects of these have largely run their course. Future gains in utilization will depend upon reducing unemployment, encouraging immigration of labour and fostering higher labour force participation at all adult ages. Increases in labour force participation are a means of addressing labour supply concerns in many countries. Japan is an exception because the labour force participation rates of men are already high and the country's family system is not conducive to retaining or recruiting married women with children into the labour force (McDonald and Kippen 2001: 23).

The presence of concurrent risks or problems detracts from the ability of societies to muster resources to address population aging. Two such problems are government debt and unemployment. Both are at high levels in many societies and their distribution overrides the broad pattern from the risk analysis. Thus most countries face economic concerns which restrict social spending options even in those with high GDP per capita. Government debt and the cost of servicing loans have long been major obstacles to improving living standards in developing countries, as well as in some developed countries. In the wake of the GFC, government debt grew. For example, the OECD's forecast of gross government debt in 2011, measured by gross financial liabilities, was 196% of 2008 GDP in Japan and 132% in Italy. The forecasts for France, the United Kingdom and the United States were also high at around 100% of the 2008 GDP (OECD 2010: 274–275). Levels of debt are subject to substantial changes through time, but these figures illustrate a widespread obstacle to meeting the extra costs of aging societies.

Unemployment is sometimes described as a greater economic problem than aging, but a combination of the two could become increasingly unsupportable. It is possible that labour shortages in populations with negative population momentum will curtail unemployment, but this would be small compensation for acquiring an even more intractable problem. Also, the early retirement of older workers in parts of Europe has not freed up expected opportunities for the young, whose unemployment rates have remained particularly high (Tobback: 2005). Before the GFC, youth unemployment rates (ages 15–24 in 2006) were at or below 10% only in group 1 countries (except Canada), in some countries from group 2 (Switzerland, Netherlands and Denmark) and in Austria and Japan. Elsewhere, the rates were above 10%, and sometimes higher than 20%, for instance in Italy, Greece and France.

17.7.2 Human Capital

The composition of cohorts entering the labour force ages is a key influence on labour utilization and productivity. The entry of young cohorts with training in new or sophisticated skills is especially significant in the technological advances that drive productivity (McDonald and Kippen 2001: 4–5). Declining cohort size in such groups can therefore have consequences for the economy disproportionate to the numbers involved. Yet enhancing human capital in other age groups – through improvements in levels of education, training and health – is also important in raising participation and productivity in the labour force as a whole. One indicator of education levels in the labour force is the proportion of people aged 25–64 who have completed a tertiary education qualification, including academic degrees and vocational courses (OECD 2010: 184). Although the data are again incomplete, especially for Eastern Europe and the developing countries, they are closer to the regional pattern in the risk analysis. In the two low risk groups (1 and 2), together with Japan, the percentages with tertiary education are mainly in the range of 30–40%, whereas in Central, Southern and Eastern European countries the figures are mostly under 25%. Russia's tertiary education figure, however, was nearly double the OECD average of 27.5%. Despite this, Russia's labour productivity was low, which UNESCO (2005) attributed to poor technology, lack of technical skills in the workforce, and low expenditure on research and development. The regional differences from the risk analysis were only partly maintained for the second measure of human capital, namely healthy life expectancy, as discussed in Chap. 5. The highest figures were for Japan, followed by Western countries, including the Southern European countries (groups 1–3). Eastern Europe and the developing countries had the lowest figures. Of the 6 year life expectancy gap between the countries of Eastern and Western in Europe, more than half is due to cardiovascular disease. The gap grew from around the late 1960s (Marmot and Wilkinson 1999: 106).

17.7.3 Policies

Resilience building to modify and ameliorate societal transformation will depend further on national approaches to policy making. Important factors will be whether governments and their constituents are accepting or unaccepting of the need for reform, whether policy approaches are proactive or reactive only to immediate circumstances, whether international co-operation is forthcoming to assist in addressing adverse national situations, and whether living standards and life styles become more consistent with prolonging life in good health. Policy responses to population aging, such as those in the *Madrid International Plan of Action on Ageing* and the OECD's principles to guide reform (Box 17.1), can mitigate some of aging's effects but cannot alter its course. Migration policies are the only widely advocated demographic responses to population aging, especially to augment numbers in the labour force. In the period 2000–2005 most Western countries had annual net migration rates of 2% or more. However, the countries least disposed to migration included some with the lowest fertility and steepest declines in their labour forces. Eastern European countries and Japan mostly had very low or negative migration rates. If their population aging remains unchecked it is conceivable that associated economic problems could prompt further labour force depletion through emigration.

Box 17.1 OECD Principles to Guide Reform

1. Public pension systems, taxation systems and social transfer programmes should be reformed to remove financial incentives to early retirement, and financial disincentives to later retirement.
2. A variety of reforms will be needed to ensure that more job opportunities are available for older workers and that they are equipped with the necessary skill and competence to take them.
3. Fiscal consolidation should be pursued, and public debt burdens should be reduced. This could involve phased reductions in public pension benefits and anticipatory hikes in contribution rates.
4. Retirement income should be provided by a mix of tax-and-transfer systems, advance-funded systems, private savings and earnings. The objective is risk diversification, a better balance of burden-sharing between generations, and to give individuals more flexibility over their retirement decision.
5. In health and long-term care, there should be a greater focus on cost-effectiveness. Medical expenditure and research should be increasingly directed to ways of reducing physical dependence, and explicit policies for providing care to frail older people should be developed.
6. The development of advance-funded pension systems should go hand-in-hand with that of the financial market infrastructure, including the establishment of a modern and effective regulatory framework.
7. Strategic frameworks should be put in place at the national level now in order to harmonise these ageing reforms over time, and to ensure adequate attention to implementation and the build-up of public understanding and support.

Source: OECD (1998: 2).

Policies supporting family welfare will be pivotal in meeting challenges of population aging in all countries and could be one of the few means of building resilience in developing countries most disadvantaged in relation to the growth of the aged population. In developed countries, family related policies are the only policy instruments that can potentially change the course of population aging and prevent it from reaching the high levels projected for Japan and countries in Southern and Eastern Europe. A long-standing generalization has been that high income countries have low fertility. Now, among the world's oldest populations, it is the high income countries in groups 1 and 2 (except Canada) that have the higher fertility rates – at least sufficient to avoid extreme levels of aging. In Europe, the lower income countries face multiple disadvantages including pressing economic problems. The map of Europe plotted the group membership of the European countries in the data set and showed regional similarities in the distribution of risks (Fig. 17.1). Because the experience of population aging is closely tied to trends in family life and fertility, accounts of these are important in explaining the regional pattern.

One such account derives from a modernization perspective on fertility and family formation in Europe (Pinnelli et al. 2001). This was based on analyses of data for 29 European countries, incorporating ideas from the second demographic transition. The study highlighted north–south and east–west contrasts within Europe. It found that higher fertility was associated with "modernisation, economic development, a more balanced gender system and greater institutional support to the family, working women, children and old people" (ibid.: 50–51). Conversely, low fertility in Central, Eastern and Southern Europe was linked with "politico-economic difficulties and a lack of modernisation", such as in relation to family policies and gender equity. The authors contrasted Southern Europe, where "state support of the family is scarce" with Nordic countries "where such support is the highest" (ibid.: 83–84). The study emphasised the fundamental importance of better economic conditions, family policies and gender equality for higher fertility:

> Higher fertility, even if below the replacement level, can be part of the framework of modernisation, in spite of new family behaviours: the framework of a more modern society, at an advanced stage of economic development, in which post-materialist values are common and more value is given to individual self-fulfilment, and the gender system is fairer, may contain not only the diversification of forms of union and their greater instability, but also fertility closer to replacement level, which is the level of fertility indicated as expected or ideal by most people in opinion surveys. In this framework, it is not "modern" to have very low fertility. The latter, if anything, is the result of difficulties which are so great as to impede marriage (or other forms of union) and fertility. (ibid.: 183).

Thus, policies that support family welfare and favourably influence the work/family balance go far towards explaining differences in fertility and population aging. The policies, in turn reflect the nature of preferred models of the family in different regions. Comparisons with the results of research by McDonald and Moyle (2010) confirms that the spatial distribution of family policy regimes broadly differentiates between the low risk profiles of France, the Nordic countries and the English-speaking countries, and the higher risk profiles of Central and Southern Europe. The higher risk countries tend to favour the male breadwinner model of the family while the lower risk countries support higher levels of gender equity in the family and the workplace. English-speaking countries have varying combinations of demographic factors underlying their relatively high fertility, but the net result seems to be that they have fewer one-child families and more families with three or more children (ibid.: 256). Although the family support services of English-speaking countries are not as substantial as those of the Nordic countries and France, taxation levels are lower and there is a fundamental value orientation in favour of the employment of mothers and a balanced combination of work and family. Low income couples have relatively high fertility rates and they receive high child-related payments. Among the English-speaking countries differences in ethnic composition probably contribute to the higher fertility of the United States (Hispanics and non-Hispanic blacks) and New Zealand (Maori and Pacific Islanders), and the lower fertility of Canada where a large proportion of immigrants are from Asia (ibid.: 270–271).

Thévenon's (2011) research on family policies in OECD countries produced somewhat similar findings although his analysis mainly derived from data on support for working parents with children under 3, and on the generosity of leave entitlements (ibid.: 64). He noted that modernization of family policies includes not only improvement of work/life balance but also advancement of a broad range of other objectives. These include poverty reduction, compensation for the economic cost of children, fostering of employment, support for early childhood development and raising birth rates – although higher fertility is more a positive potential by-product rather than an explicit objective. Thévenon's analysis separated the Nordic countries from the others in the two low risk groups (groups 1 and 2). He found that policies in the Nordic countries, including Sweden, Norway and Denmark, were the most family friendly, providing strong support for working parents with children under 3 (ibid.: 65). The policies of these countries include "state-subsidized child care, paid parental leave, family carers leave, regulations that restrict hours of work and income transfers through the tax or transfer systems that support families with children" (McDonald and Moyle 2010: 248). In Thévenon's study, other countries in Western and Central Europe provided relatively high levels of cash support to families through benefits and tax breaks, but this was marked by conservatism in relation to gender equity and the reconciliation of work and family life. For example he found that in France and Germany traditional one-earner households had lower tax rates than two-earner families. In France, however, this was offset by other advantages such as a home care allowance and provision of child care (Thévenon 2011: 72 and 77). Thévenon found that in Southern and Eastern Europe and Japan public support for families was quite limited and highly fragmented: working hours were long and there was a lack of formal childcare support (ibid.: 77).

It is uncertain whether family policy initiatives could produce an increase in fertility in regions of Europe with the lowest fertility, because this could depend on changes in priorities and values throughout whole societies, as well as on the availability of substantial resources for implementation. Yet it is apparent from the experience of other regions of Europe and English-speaking countries that family friendly policies can have a positive impact on fertility, although their introduction may require society-wide reform and a full range of incentives and supports (McDonald 2006: 500–505). The most straight-forward suggestion for policy reform to foster higher average completed family size and sustainable population aging is through provision of high quality, worker-friendly, affordable child care. The underlying hypothesis is that: "As child care becomes more available, affordable, and acceptable, the antinatalist effects of increased female educational attainment and work opportunities decrease" (Rindfuss et al. 2010: 725). From a study of Norway, the authors concluded that having places available for 60% of pre-school children leads to an increase of between 0.5 and 0.7 in the average number of children per woman. Although this finding originated in a small prosperous country where, since the 1980s, a range of policies have sought to promote gender equity in the family and the workplace, the authors argued that it could be applied in Japan. They considered that child care provision would lead to higher fertility even without any change in gender equity. They also noted that Germany is pursuing this strategy, in conjunction with other policy changes to better reconcile work and family life for women. Hitherto West Germany in particular had low availability of childcare places and there was much reliance instead on grandparents and other relatives (ibid.: 742–743).

17.8 Conclusion

Adequate GDP per capita, high levels of human capital, and prudent development of policies in relation to migration and family welfare are important in building resilience as population aging occurs. The analysis of risks reveals clear distinctions between groups of countries with relatively low and relatively high risks. This division corresponds reasonably well with the distribution of countries in terms of resilience, especially in the association between low risks and high resilience. A number of countries with high risks and at least some positive indicators of resilience evidence potential to modify their unfavourable demographic trends.

As discussed in previous chapters, the welfare of individuals, families and communities also depends on the extent to which the needs of older people can be met through a wide range of other social policies which international organizations have debated, clarified and advocated. The countries best placed to implement these most fully are again those with low risks and advantages in relation to societal resources. No country can expect business as usual as its population ages, but timely initiatives can prevent crises. Some of the main impediments to an ordered transition to an older age structure are rapid population growth or decline, high levels of government debt and other long-term economic and environmental problems that limit responses to population aging. The most uncertain and troubled futures in relation to population aging confront countries with high risks and low resilience.

The transformative effects of population aging are evident at all levels of society, such as in relation to individual life course experience and longevity, family development, community change, the labour force, and health and welfare systems. Population aging has resulted from human decisions but it has acquired an impetus of its own. The more it is neglected, the less amenable it will be to interventions to address its previously unexpected and unintended consequences. A beneficial consequence, however, has been the growth in the numbers of healthy, active older people who are ready and able to contribute to society. There has been only slow realization of this and many obstacles to their participation remain, including underestimation of their potential and barriers to working and earning in later years to the extent that individuals are disposed and capable. Universal characteristics of older people need to include social integration and, especially in the Third Age, access to opportunities for productive engagement.

References

Cai, Y. (2010). China's below-replacement fertility: Government policy or socioeconomic development? *Population and Development Review, 36*(3), 419–440.

CIA. (2010). *The world factbook*, country comparison: GDP per capita (ppp). https://www.cia.gov/library/publications/the-world-factbook/rankorder/2004rank.html. Accessed Sept 2010.

Marmot, M., & Wilkinson, R. G. (Eds.). (1999). *Social determinants of health*. Oxford: Oxford University Press.

McDonald, P. (2006). Low fertility and the state: The efficacy of policy. *Population and Development Review, 32*(3), 485–510.

McDonald, P., & Kippen, R. (2001). Labor supply prospects in 16 developed countries, 2000–2050. *Population and Development Review, 27*, 1–32.

McDonald, P., & Moyle, H. (2010). Why do English-speaking countries have relatively high fertility? *Journal of Population Research, 27*(4), 247–273.

OECD. (1998, June). *Maintaining prosperity in an ageing society: Summary.* Paris: Organisation for Economic Co-operation and Development. http://www.oecd.org/dataoecd/21/10/2430300.pdf. Accessed Sept 2010.

OECD. (2009). *StatExtracts*. Paris: Organisation for Economic Co-operation and Development. http://stats.oecd.org/Index.aspx?DataSetCode=DECOMP. Accessed Sept 2010.

OECD. (2010). *Factbook 2010: Economic, environmental and social statistics*. Paris: Organisation for Economic Co-operation and Development.

Pinnelli, A., Hoffmann-Nowotny, H. J., & Fux, B. (2001). *Fertility and new types of households and family formation in Europe* (Population studies no. 35). Strasbourg: Council of Europe Publishing.

Population Reference Bureau. (2008). *2008 world population data sheet.* Washington, DC: Population Reference Bureau.

Rindfuss, R. R., Guilkey, D. K., Morgan, S. P., & Kravdal, Ø. (2010). Child-care availability and fertility in Norway. *Population and Development Review, 36*, 725–748.

Swan, W. (2010). *Australia to 2050: Future challenges (the 2010 intergenerational report).* Canberra: Commonwealth of Australia.

Thévenon, O. (2011). Family policies in OECD countries: A comparative analysis. *Population and Development Review, 37*(1), 57–87.

Tien, H. Y., et al. (1992). China's demographic dilemmas. *Population Bulletin, 47*(1, June), 1–44.

Tobback, B. (2005, October 17–18). Chair's conclusions. *High-level forum on ageing and employment policies*, Brussels. www.oecd.org/olderworkersforum. Accessed Apr 2010.

UNESCO. (2005). *Education trends in perspective: Analysis of the world education indicators.* Paris: UNESCO Institute for Statistics. http://www.uis.unesco.org/TEMPLATE/pdf/wei/WEI2005.pdf. Accessed Sept 2010.

United Nations. (2004). *Policy responses to population decline and ageing* (Population bulletin of the United Nations, special issue nos. 44/45). New York: United Nations.

United Nations (2009). *World population prospects: The 2008 revision population database.* New York: Population Division of the Department of Economic and Social Affairs. http://esa.un.org/UNPP/index.asp?panel=2. Accessed Apr 2010.

Walker, A. (1999). *Attitudes to population ageing in Europe: A comparison of the 1992 and 1999 Eurobarometer surveys.* European Commission: http://ec.europa.eu/public_opinion/archives/ebs/ebs_129_en.pdf. Accessed Feb 2011.

WHO. (2009). *Statistical information system.* http://www.who.int/whosis/data/Search.jsp. Accessed June 2011.

World Bank. (2011). *World development indicators, databank: Central government debt, total (% of GDP).* http://data.worldbank.org/indicator/GC.DOD.TOTL.GD.ZS/countries. Accessed Apr 2011.

Index

D.T. Rowland, *Population Aging: The Transformation of Societies*, International Perspectives on Aging 3, DOI 10.1007/978-94-007-4050-1,

D

E